REGRESSION
BASICS

REGRESSION
BASICS

LEO H. KAHANE

Sage Publications
International Educational and Professional Publisher
Thousand Oaks ▪ London ▪ New Delhi

For information:

Sage Publications, Inc.
2455 Teller Road
Thousand Oaks, California 91320
E-mail: order@sagepub.com

Sage Publications Ltd.
6 Bonhill Street
London EC2A 4PU
United Kingdom

Sage Publications India Pvt. Ltd.
M-32 Market
Greater Kailash I
New Delhi 110 048 India

Printed in the United States of America

Library of Congress Cataloging-in-Publication Data

Kahane, Leo H.
   Regression basics / by Leo H. Kahane.
      p. cm.
   Includes bibliographical references and index.
   ISBN 0-7619-1958-9 (cloth: acid-free paper) —
   ISBN 0-7619-2413-2 (pbk.: acid-free paper)
   1. Regression analysis. I. Title.
   QA278.2 .K34 2000
   519.5'36—dc21   00-013066

01   02   03   10   9   8   7   6   5   4   3   2   1

Acquiring Editor:	C. Deborah Laughton
Editorial Assistant:	Eileen Carr
Production Editor:	Diana E. Axelsen
Editorial Assistant:	Candice Crosetti
Typesetter/Designer:	Technical Typesetting, Inc.
Cover Designer:	Michelle Lee

CONTENTS

ACKNOWLEDGMENTS

I wish to thank C. Deborah Laughton who was a great source of encouragement on this project from beginning to end. I also wish to thank Jagdish Agrawal, who diligently read earlier drafts of this book and whose comments ultimately led to significant improvements. Finally, I wish to thank Terry Smith and Elisabeth Cisse for help with the art work.

To Cathy and Jake, my special prizes.

PREFACE

Regression analysis has become an integral, regularly used research tool across a wide variety of disciplines, including political science, education, sociology, and economics, just to name a few. In fact, a large number of people (some unknowingly) have personal computers with software capable of performing regression techniques. Thus, with a few quick keystrokes almost anyone can perform a regression analysis, and many do. However, the key is to know what one is doing with this powerful tool, and why. Indeed, a colleague of mine used to say that personal computers have done to regression analysis what microwave ovens have done to cooking: Nearly everyone has a microwave and by popping a frozen chicken cordon bleu dinner into the oven, people think they are gourmet chefs. This is where this book comes in.[1] The purpose of this book is to provide an intuitive, nontechnical introduction to the topic of regression analysis.[2]

I have several key goals in mind with this book. First, it is designed to provide an introduction to regression analysis for individuals seeking to understand the basic ideas of the subject. Second, as a teacher of regression analysis, I find that students unfamiliar with the topic find it difficult to simply pick up a textbook and begin reading with understanding. Thus, this book is intended to be a companion to a more comprehensive textbook used in a course on regression analysis. An introductory-level understanding of statistics is the only prerequisite for the materials presented in this book.

A third goal of this book is to demonstrate the flexibility and wide application of regression analysis in research. It is often thought by those not acquainted with the subject that regression analysis is primarily a tool used to forecast financial variables. Although this is one of its uses, regression analysis can be used in many, often quite creative ways. In this book, for example, we will use regression analysis to determine the factors explaining salaries in professional sports and learn why some players earn more than others. In the sphere of politics, we will consider the factors that affect voting patterns in presidential elections. And, in a third example, we will consider the sensitive, social issue of abortion and determine the important factors explaining the difference in abortion rates across the United States. These three examples will be studied in the chapters ahead and, as we shall see, the tools of regression analysis can be used to explore these phenomena and many others.

The way this book is organized is to start with the simplest (two-variable linear) model and work toward the more complex multivariate regression model in later chapters. In all chapters, the discussion will use the examples mentioned previously to illustrate and motivate the ideas at hand. The data for these examples are provided in Appendix A. In addition, two other examples, one that studies the effects of education (as well as other factors) on a person's salary and another that studies the determining factors of automobile prices, are presented as problems at the end of each chapter. These data are also provided in Appendix A, and all data are available on line.[3] It is *highly recommended* that the reader use the data provided in Appendix A to replicate the results found in the text.[4] In doing so, the reader can have tangible, hands-on experience in *performing* linear regression analysis. Indeed, this is the ultimate goal of this book: to equip the reader with the knowledge and ability to perform and interpret the results of basic linear regression analysis. As noted, there are problems at the end of Chapters 1 through 6, and the answers to these problems are provided in Appendix D. The reader is encouraged to work through these problems as they should help the reader have a better understanding of the concepts presented in this book. Finally, throughout the book, there are words or terms that are printed in boldface type. These are key terms whose definitions appear in the Glossary at the back of this book.

———•◆•———

▼ Notes

1. I don't wish to imply that one should "cook the data!" The microwave analogy is attributed to friend and colleague, Andy Abere.

2. Regression analysis is the core subject in a field of study that economists call **econometrics**. In various points throughout this book the reader is referred to books on econometrics for further information.

3. The data can be found (as ASCII files) at the following Web site: www.sbeusers.csuhayward.edu/~lkahane/index.html.

4. There are many computer software programs capable of performing regression analysis. Examples of software that is simple to use, but with limited capabilities include Microsoft Excel and Lotus 1-2-3. More complex, yet comprehensive software includes SAS and SPSS. The analyses performed in this book can be carried out with Microsoft Excel and SPSS.

AN INTRODUCTION TO THE LINEAR REGRESSION MODEL

The basic goal of regression analysis is to use data to analyze relationships. Thus, the starting point for any regression analysis is to have something to analyze. That is, we begin with some idea or "hypothesis" we want to test and we then gather data and analyze them to see if our idea is verified. The purpose of this chapter is to provide the reader with several examples of the kind of research that can be done with regression analysis techniques. These examples, which are woven throughout this book, were chosen in such a way as to illustrate to the reader how regression analysis methods can be used to understand relationships across a broad range of subjects. Once we understand the basic notion of regression analysis, we then proceed to Chapter 2 where the more technical aspects of regression analysis are discussed.

INTRODUCTION

Baseball Salaries

Suppose we are interested in exploring the factors that determine one's salary. There are many such factors, one of which would be the experience an individual has in his or her profession. That is, for most professions, the longer a person has been on the job, the greater is his or her salary. The logic behind this relationship is that workers learn with experience and become more productive over time. As such, employers reward workers for their increased productivity that comes with experience. But how large is the reward to experience? That is, as a person gains another year of work experience, by how much can he or she expect his or her salary to increase from one year to the next? One method of trying to understand this relationship between salary and experience is to collect data on individuals within a profession and use a graph to visualize the relationship between the two. As an example, let's consider the occupation of professional baseball players.

Salaries earned by Major League Baseball (MLB) players have been the subject of great discussion in the media largely because in recent years players have earned enormous amounts of money for playing the game. We may consider, then, how a player's salary is related to his experience in MLB.[1] Data on players' salaries have become public information these days as a number of media sources publish the earnings of players as well as other information about them, such as years of MLB experience.[2] Suppose we collected a sample of data on player salary and experience and plotted these pairs of numbers on a graph with a player's salary on the vertical, or Y, axis and the corresponding years of experience on the horizontal, or X, axis. Figure 1.1 shows an example of how this graph may look.

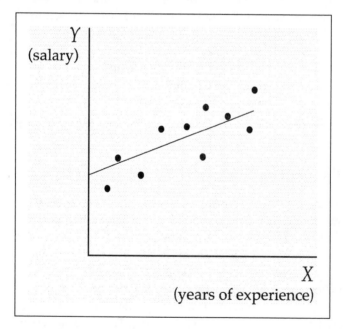

Figure 1.1.

Viewing Figure 1.1, we can observe that the collection of dots, each of which represents an individual player's salary and his associated experience, tends to rise as we move out along the X axis. As a means of trying to represent the general behavior of these dots, a line has been run through them that shows their general tendency to rise. As the line suggests, as a player's experience (X) increases, his pay (Y) tends to increase as well. This would seem to support our hypothesis that workers (players) are rewarded with greater salaries as their experience (years of playing MLB) increases.

By adding a line to Figure 1.1, we were able to capture the general relationship between salary and experience. In doing so, however, it also implies a more specific assumption about the behavior of Y with respect to X. This assumption, known as the linear regression model assumption, forms the basis for regression analysis and is explained in the following section.

Linear Regression Model Assumption

The easiest way to understand the linear regression model assumption is to illustrate it with an example. Returning to our case of baseball, suppose instead of just a sample of data, we collect data for *all* MLB players. Having such a large collection of data, we could then order our data such that players are grouped according to the number of years of MLB experience, which was our X variable for this example. Thus, all players with, say, 1 year of experience would be grouped together. All players with 2 years of experience would be in another group. And so on. We could then record the salary of each player in each group, and then use this information to calculate the average salary for each group as well. This procedure is illustrated graphically in Figure 1.2(a).

Viewing Figure 1.2(a), we can consider players with 1 year of MLB experience who have their salary plotted on the graph above the value shown as 1 on the X axis. Notice that some players in this group have higher salaries than others, perhaps because of differences in other skills (this point is expanded on later). If we calculated the average salary of players in this group, its value would lie somewhere in the middle of these plotted points, such as the point shown with a heavier dot. Thus, this heavy dot represents the mean or average salary of players with 1 year of experience. We can carry out this same exercise for

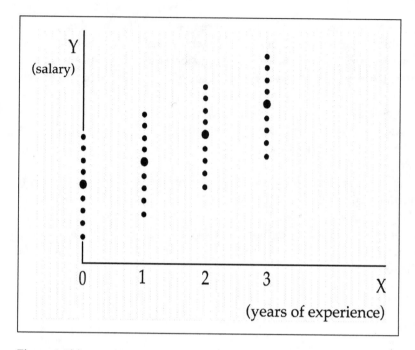

Figure 1.2(a).

players with 2 years of MLB experience. These individuals have their salary plotted above the value of 2 on the X axis. As in the previous case, some players in this group will have higher salaries than others, and the average salary for all players with 2 years of MLB experience is shown by the heavy dot above 2 on the X axis. This same kind of analysis can be done for players with 3 years of MLB experience, and the heavy dot above the value of 3 on the X axis represents the mean salary for all players in this group. This procedure can, in fact, be done for all values of X, MLB experience, in each case calculating the mean value for salary (Y) for given values of experience (X). Given this graph, then we have the following assumption: *The linear regression model assumes that the mean values of Y, for given values of X, is a linear function of X.* Or, in terms of our graph, the heavy dots (which are the mean values of Y for given values of X) lie on a line. (It should be noted that, in some cases, the relationship between the mean values of Y and X may be *nonlinear*. Examples of nonlinear relationships are discussed in Chapter 5.)

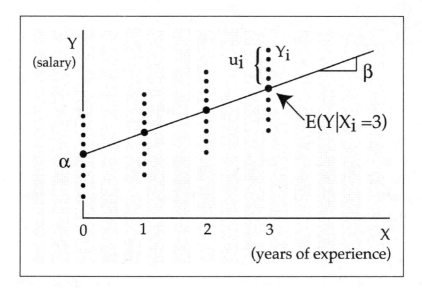

Figure 1.2(b).

This assumption is shown graphically in Figure 1.2(b), which takes Figure 1.2(a) and adds a line connecting the heavy dots.

This assumption can also be expressed somewhat more formally by using the following mathematical expression:

$$E(Y|X_i) = \alpha + \beta X_i. \tag{1.1}$$

The E in (1.1) stands for "expected value" or mean, and the vertical line, |, can be read as "for given values" of X_i. (The subscript i is used to keep track of different values that X can take on.) The expression on the right-hand side of the equal sign, $\alpha + \beta X_i$, is simply the equation to a line. Or, putting it all together, (1.1) can be read as: "the expected value of Y for given values of X_i is equal to a linear function of X_i." As for terminology, the variable Y is called the **dependent variable** as its value is said to depend on the value that X_i, which is called the **independent variable**, takes on.[3] The symbols α and β in (1.1) are constants and are referred to as the intercept and slope term, respectively (it is common in regression analysis to use Greek letters such as these). The intercept term α tells us what the expected value of Y (in our case, salary) would be for individuals who have no experience (i.e., new, or "rookie," MLB players). That is, if a player has no experience, then his value for X_i is 0. Plugging in 0 for X_i in (1.1), we have

$$E(Y|X_i = 0) = \alpha + \beta(0) = \alpha. \tag{1.1a}$$

This is shown in Figure 1.2(b) where the line representing (1.1) intersects the vertical axis.

The slope term β in (1.1) tells us how Y changes for each one-unit increase in X_i. Or, in the case of MLB salaries, how salaries increase for each additional year of experience. To see this, consider a player with 1 year of MLB experience. His value of X_i, then, is 1, and plugging this into (1.1) yields

$$E(Y|X_i = 1) = \alpha + \beta(1) = \alpha + \beta. \tag{1.1b}$$

Comparing (1.1a) and (1.1b), we see that the difference between the two is that players with 1 year of experience are expected to have β more in salary than players with no experience. In the case of players with 2 years of experience, they are expected to have 2β more in salary as compared to those with no experience. Thus, each additional year's worth of experience increases a player's expected salary by β.

The implication of writing the equation with $E(Y|X_i)$ is that it implies the understanding that *individual* values for Y, for given values of X_i, will not likely be exactly equal to $\alpha + \beta X_i$. To see this, we can return to Figure 1.2(b) and consider player i who has 3 years of experience in MLB and earns a salary of Y_i. Notice that, for this individual player, his actual salary Y_i is greater than the mean salary for his group shown as $E(Y|X_i = 3)$. The difference between the actual and expected value for Y is shown as u_i. In terms of an equation, we can write a player's actual salary as

$$Y_i = E(Y|X_i) + u_i. \tag{1.2a}$$

Or, using (1.1), we can rewrite the last equation by replacing $E(Y|X_i)$, giving us

$$Y_i = \alpha + \beta X_i + u_i. \tag{1.2b}$$

Every dot shown in Figure 1.2(b), which represents a particular player's experience and his actual salary, can be expressed in a similar way. That is, every individual player's salary can be expressed as the sum of his group's expected salary, plus the specific player's value for u_i. What does u_i represent? The term u_i, which is called the **error term**, represents all the other factors that may affect player i's salary, which are not taken into account by our simple model shown in (1.1). There are, in fact, numerous other factors that enter into the determination of salaries. In baseball, for example, players are rewarded for their offensive (e.g., hitting) and defensive (e.g., fielding) abilities. The fact that these other important explanatory variables are not accounted for in our model means that player salaries would not likely fall exactly on the line shown in Figure 1.2(b). To further illustrate why this is the case, consider two players who are identical in all measures, including years of experience, except one player is a better hitter. This being the case, the better hitter would likely earn a greater salary because he is worth more to a team. What this means, then, is that although a player's experience may be an important factor in explaining his salary, experience alone cannot perfectly explain a player's salary. The error term included in (1.2b) is said to be **stochastic**, meaning that it is a random component of a player's salary that varies from one player to another. Thus, if we again consider our specific player i who has $X_i = 3$ years of playing experience, as shown in Figure 1.2(b) the vertical distance from the heavy dot on the line to the point representing this player's salary is the positive error u_i. This means that player i is paid more than expected, perhaps because he is a better hitter, a factor not taken into account in our simple model. In a similar way, points below the line represent players whose salaries are less than expected (i.e., they have negative errors), perhaps because they are below-average hitters.

At this point, the reader may be wondering if it is possible to build a more elaborate model that takes into account these other factors that are missing from our model and that end up in the error term, u_i. To a certain extent, this can and will be done in later chapters when we build on this simple model to include other explanatory measures such as hitting and fielding. In any case, it is not likely that *all* factors can be accounted for so that the error term is driven to 0.[4] This is true for a number of reasons. First, there may not be data available for many important variables (e.g., a player's speed in running the bases). Second, some factors that affect a player's salary may not be measurable (e.g., leadership ability or fan appeal). All these factors that are not accounted for in our model end up in the error term, which will vary from player to player.

For now, we will continue to work with simple models like that shown in (1.2b), which are referred to as **two-variable linear regression models** (also know as **bivariate linear regression models**) because they include only an intercept (α) and one slope term (β). These simple models will serve as a starting point from which we can discuss many of the issues regarding regression analysis. Bear in mind, though, that in most cases a two-variable model will be too simplistic for our purposes and a more complex model will be needed.

Population Data Versus Sample Data

Before moving on, we need to clarify some aspects of our data sets, namely, their size. Typically, when we consider a theory, such as MLB salaries as a function of years of experience, there is a relevant population of data. In the baseball example, it may be all MLB players, past and present. For this population of data, when we formulate a mathematical model for the behavior of a dependent variable as a function of an independent variable, we are constructing what is called the **population regression function** (PRF) because it presents a hypothesis about the behavior of the population of data. Thus, for MLB, the model shown in (1.1) is a population regression function for salary determination in MLB. In most cases, however, it is not possible to collect data for the entire population, perhaps because the data do not exist or because it would be practically impossible to collect the data.[5] As such, samples of data are collected from the population and analyzed with the hope that the information contained in the sample is a good representation of the how the population behaves. To keep the distinction between sample analysis and population analysis clear, we will use the following **sample regression function** (SRF):

$$\widehat{Y}_i = a + bX_i, \tag{1.3}$$

where $\widehat{Y}_i$ is the sample version of the expression $E(Y|X_i)$ and a and b are the sample versions of the population's α and β. Figure 1.3 shows a graph of the sample regression function. As in the case for the population, given our sample, we can represent a specific player i's salary as the sum of what our model predicts his salary to be based on his experience, plus the error in prediction, e_i:

$$Y_i = \widehat{Y}_i + e_i. \tag{1.4}$$

Using (1.3), we can rewrite this expression by substituting for $\widehat{Y}_i$ giving us

$$Y_i = a + bX_i + e_i. \tag{1.5}$$

Thus, (1.5) shows the linear relationship between Y_i and X_i, with the term e_i representing all other factors not accounted for in our model. This equation will be used in place of (1.2b), which was for the population data, and the intercept term a is a sample estimate of the population's α and the slope term b is a sample estimate of the population's β. The term e_i is the error term (also called the **residual**) for our sample regression function and is analogous to the population's error term u_i.[6] As shown in Figure 1.3, the residual is simply the difference between

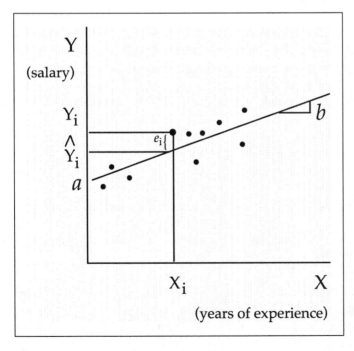

Figure 1.3.

player i's actual salary, Y_i, and the salary we would predict for a player with X_i years of experience, $\widehat{Y}_i$ (i.e., the point on the line above X_i).

It is hoped that the sample's intercept and slope terms closely resemble the population's parameters α and β. If this is the case, then we can be confident that by analyzing the sample's values for these parameters we can understand the behavior of the population.

Presidential Elections

As a second example of a regression analysis model, we can consider the topic of presidential elections. Some academics, such as Yale economist Ray C. Fair, argue that the state of the economy is an important factor in describing the voting pattern in presidential elections.[7] As Fair puts it, "...voters hold the party in the White House responsible for the state of the economy."[8] For example, if the current president is a Democrat, and the economy has grown substantially during his term, then the party in power is given partial credit for that economic success and voters would then reward the Democratic presidential candidate with votes. On the other hand, if the economy has suffered from recession in the years prior to the election, the reverse is true and the incumbent-party candidate suffers. This theory can be evaluated using regression analysis. We can model voting for incumbent-party candidates with the following sample regression function:

$$Y_t = a + bX_t + e_t. \tag{1.6}$$

In (1.6), we now have the dependent variable, Y_t, representing the percentage of the two-party votes received by the candidate running for president who belongs to the same party as the incumbent (note that this could be the incumbent himself if he is running for a second term, e.g., Ronald Reagan who ran for president in 1984 and who was president in 1980). The variable X_t now represents the economy's real percentage growth rate over some specified period prior to the election at hand.[9] In this case, the error term, e_t, represents other factors not taken into account, such as the inflation rate prior to the election and perhaps other, immeasurable factors such as the charisma of the candidate. Finally, note that, in this case, we use the subscript t (as opposed to i used for the baseball example) to distinguish individual cases because now we are considering results of elections at different points in time. Graphically, this model would look similar to the one shown in Figure 1.3, except the values for X_t, the real growth rate of the economy, can be negative. This is shown in Figure 1.4, which has a positive and negative range for X_t.

The value of a in this case would be the expected share of the two-party presidential vote received by the incumbent-party candidate when the economy experienced no real growth (i.e., when $X_t = 0$). As for b, this would represent the increase (decrease) in the share of votes the incumbent-party candidate would receive for a 1 percent increase (decrease) in the real growth rate. As in the case of baseball salaries, the actual data for Y_t would not likely fall exactly on the line, but would be "speckled" above and below the line as shown in Figure 1.4. The vertical distance from these observations to the line shown would be the error term e_t, which, again, represents other factors affecting the share of the two-party votes the incumbent candidate received that are not taken into account in our simple model. As shown in Figure 1.4, for the

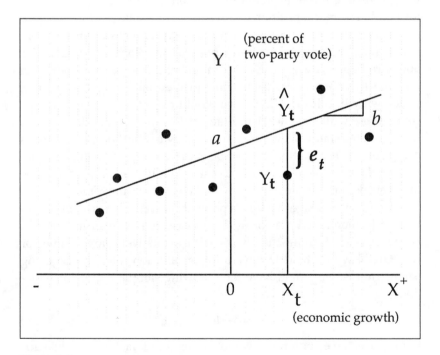

Figure 1.4.

given value of X_t the actual value of Y_t lies below the value predicted by the regression line, $\widehat{Y}_t$. Thus, the associated error term e_t, which is equal to the actual value of Y_t minus its predicted value, would be negative.

Abortion Rates

Finally, our third example of a regression analysis model deals with the social issue of abortion. As mentioned in the Preface, abortion rates (number of abortions performed per 1000 women of child-bearing age) across the United States differ, sometimes greatly. Researchers have been interested in discovering what factors play a role in explaining why in some states the abortion rate may be relatively high, whereas in others it is relatively low. There are, of course, many factors that affect the abortion rate across states, but one of them would likely be the moral views of the state's residents. Other things being equal, the greater the moral aversion to abortion, the fewer would be performed.[10] The moral position that residents of a state hold with regard to abortion is difficult to measure. One way to measure it is to consider what percentage of the state's population belongs to the Catholic, Southern Baptist, Evangelist, or Mormon faith. These are the four main religions that have a stated opposition to abortion. Using this measure, which we will call RELIGION, we would expect that, if we compare states, those with a greater percentage of state population that belongs to one of these faiths would tend to have fewer abortions, other things being equal. This relationship can be expressed, again, using an equation similar to those in our previous examples. In this case, Y_i would be the abortion rate in state i, and X_i would be the measure for RELIGION, equal to the percentage of a state's population that belongs to one of the four faiths mentioned previously. In this case, the slope term, β, would be negative, indicating that states with a relatively large value for RELIGION would tend to have a lower abortion rate, all else being equal. That is, we would have the following sample regression equation:

$$Y_i = a + bX_i + e_i. \tag{1.7}$$

In this case, the error term e_i would capture other factors omitted such as income and legal differences across the 50 U.S. states. It should be noted that in this example the subscript i is used to keep track of values for Y and X for states (not individuals, as was the case in the baseball example). Graphically, we would have something like Figure 1.5. As in the other examples, the vertical distance from a given point on the graph (e.g., Y_i, X_i) to the line would represent the error term e_i. In the example shown in Figure 1.5, the actual observation for the dependent variable, Y_i, lies below the predicted one, $\widehat{Y}_i$, for the given value of the independent variable, X_i, and so the error term would be negative.

As for the intercept, a, it requires some additional discussion in this example. Technically speaking, the intercept represents the abortion rate that we would expect in the case where the variable RELIGION is 0 (i.e., $X_i = 0$). However, intuition would tell us that it is clearly not the case that the variable RELIGION would take on the value of 0 in any state as this would require that *none* of a state's residents belongs to the Catholic, Southern Baptist, Evangelist, or Mormon faith. As we can see in Appendix A, which presents the data for the variable RELI-GION, our intuition is correct as the value of RELIGION is, in fact, not 0 in any state. What this means for our model shown in (1.7) is that, for

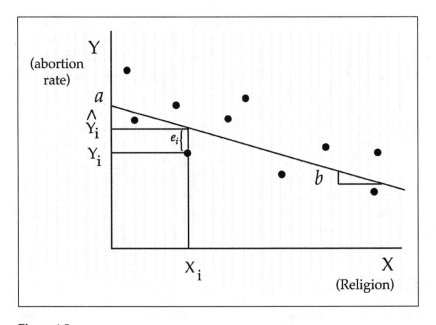

Figure 1.5.

this example, the intercept has no sensible interpretation. That is, the intercept term is technically necessary to "anchor" the line in the graph, but beyond that it is meaningless. (The fact that the intercept may have no meaningful interpretation is indeed often the case in regression model analysis.)

As for the value for the slope term, b, in this case it represents the change in the abortion rate as the measure for RELIGION increases by 1%.

Thus, we have three interesting examples that we can use to develop the basics of linear regression analysis. The graphs presented previously for our three models show a line drawn through a scatter of points. At this point, two questions arise. First, what is the ultimate use of knowing such a line? The answer to this question is that, if we know this line, it can be used to predict values of the dependent variable for given values of independent variables. For example, considering our model of MLB salaries, if we know the values (i.e., the numbers) of the intercept, a, and the slope term, b, we could then use the line defined by them to predict a player's salary given his experience. This could be useful, for example, if a player is interested in knowing what his salary is expected to be as his experience increases. We will see in the proceeding chapter how this kind of prediction can be accomplished. Of course, we must keep in mind that the predictions we make may not be entirely accurate because other factors that may be important in explaining a player's salary are not being considered. As we have already discussed, these excluded factors end up in the error term, e_i.

The second question that arises is: How can we find values for a and b so that the defined line "best fits" the scatter of points, which are our actual data? Recall that, in our three examples of sample regression functions, we simply added a line to our graphs in such a way that the line seemed to fit the data well. This visual method, however, is imprecise and there are better methods for accomplishing this task. This question of how we find the best-fitting line to the data is, in fact, the subject of the next chapter.

Types of Data Sets

Finally, before moving on to the next chapter, we must say a few things about the various types of data sets. There are essentially three general varieties: a **cross-section**, a **time series**, and a **pooled data set**. A cross-section data set fixes a point in time and looks across space. Our baseball example is a cross section as we collect data on salary and experience for a particular year and consider how salaries differ across players. In addition, our abortion example is a cross section as we consider a single point in time and look across the 50 U.S. states. A time series follows variables across time, while holding space constant. Thus, our presidential election example is a time series as we follow the breakdown of votes from one election to another. A pooled data set is a combination of both. For example, if we followed baseball salaries paid to all players *and* from year to year, we would have a pooled data set. Pooled data sets are common, but often require special treatment that goes beyond the scope of this introductory book.[11]

———•◆•———

▼ Notes

1. There has been, in fact, a great deal of empirical research done on this topic. See, for example, Scully (1974) and Zimbalist (1992).

2. For example, data on player salaries are published annually by a number of newspapers (e.g., *USA Today*) and are also available on line at various Internet sites, including one maintained by Sean Lahman entitled "The Baseball Archive," which is located at www.baseball1.com. Information on player experience and performance can also be found at this web site and is published annually in various other sources, including Thorn and Palmer's *Total Baseball*.

3. The independent variables are also sometimes referred to as "predictors" or "explanatory variables."

4. There is a case when the error term will, in fact, be 0. This is when an identity has been estimated. For example, suppose we collected data for distance measurements in meters and then collected data for the same distance measurements in inches. If we tried to estimate the relationship between meters and inches, we would find a perfect linear relationship and the errors would all be 0. This would be the result because 1 meter is defined to be exactly 39.37 inches and if measurements are made carefully enough, there should be no errors. There is no reason, however, to estimate an identity because these relationships are already known.

5. Suppose, for example, we were studying the eating habits of the U.S. population. It would be nearly impossible to collect information from every individual given the U.S. population is approximately 275 million.

6. Some authors reserve the term "residual" only for the sample regression function's error term (e_i) and use "error" or "disturbance" for the population regression function's error term (u_i). We will use both residual and error term for e_i, remembering that these refer to sample results.

7. Fair (1996). See also Kramer (1971) and Stigler (1973).

8. Fair (1996), page 90, footnote 2.

9. The "real" growth rate is a term economists use to refer to the economy's growth rate, adjusted for inflation.

10. Previous research on the determinants of abortion rates can be found in Medoff (1988) and Kahane (2000).

11. Greene (2000) provides an advanced discussion on the handling of pooled data sets.

PROBLEMS

1.1   Consider the following model:

$$Y_i = \alpha + \beta X_i + u_i,$$

where Y_i is individual i's wage and X_i is individual i's years of education.

a. What is the interpretation of α? Do you expect it to be positive or negative?

b. What is the interpretation of β? Do you expect it to be positive or negative?

c. What does the error term, u_i, capture in this case?

1.2   Consider the model for presidential elections given in (1.6), which shows the percentage of two-party votes received by the incumbent-party candidate. What other factors might be important in determining Y_t besides the real growth rate?

2

THE LEAST-SQUARES
ESTIMATION METHOD: FITTING
LINES TO DATA

In the three examples discussed in the previous chapter, lines were drawn in such a way as to best fit the data at hand. The question arises as to how we find the equation to such a line. This is the point of linear regression analysis: fitting lines to data. We can consider a number of approaches. For example, we could consider simply using a ruler and drawing a line that seems to fit the data best. This method, however, is not advisable because it is not a precise approach and in some cases the scatter of points does not suggest to the naked eye an obvious location for the line. A more systematic approach will be presented in this chapter.

INTRODUCTION

Ordinary Least-Squares

One possibility would be to find a line that would minimize the sum of the error terms. This approach, however, is flawed. To see this, suppose we have a very small sample of just four data points, which are plotted in Figure 2.1. The line shown seems to fit the data well using the criterion that the sum of the errors is minimized. In fact, if we calculate $e_1 + e_2 + e_3 + e_4$, it would come to approximately 0 as the small positive errors (e_2 and e_4) would cancel out the small negative errors (e_1 and e_3), indicating a good fit. Unfortunately, under this criterion the line shown in Figure 2.2 would serve just as well. As in the previous case, the sum of the errors here is also approximately 0 as the small positive error

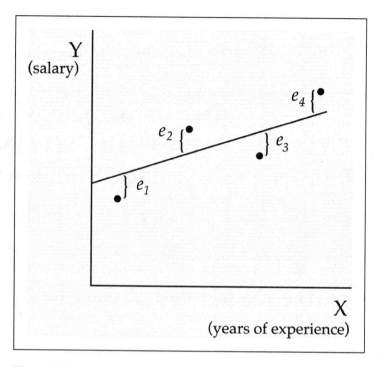

Figure 2.1.

cancels out the small negative error (e_3 with e_2) and the large positive error cancels out the large negative error (e_4 with e_1). Thus, we have two very different lines that meet our criterion of minimizing the sum of the errors equally as well. In fact, there are any number of such lines that when drawn would give us a zero sum of errors. Obviously, this criterion will not do.

The problem with the last method considered is the cancellation of positive errors with negative errors. A way of avoiding this problem would be to find a line such that the sum of the *squared* errors is minimized.[1] That is, our task is to find a line determined by a and b such that the sum of the squared errors, $e_1^2 + e_2^2 + e_3^2 + e_4^2$, is as small as possible. This, in fact, is called the method of "least-squares," or sometimes called **ordinary least-squares (OLS)** so as to distinguish it from some other specialized least-squares methods.[2] We can represent this task of minimizing the sum of squared errors mathematically by first noting

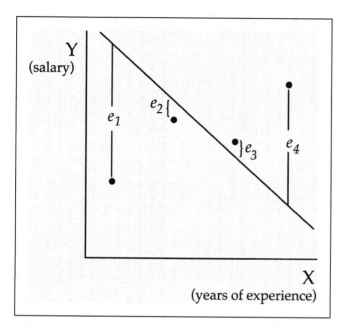

Figure 2.2.

that the error term e_i can be rewritten, using (1.5), in the following way:

$$e_i = Y_i - (a + bX_i).$$ (2.1a)

Thus, the error is decomposed into the difference between the observed value Y_i, and the predicted value, $(a + bX_i)$. Or, getting rid of the parentheses, we have

$$e_i = Y_i - a - bX_i.$$ (2.1b)

The OLS method finds a value for a and b that minimizes the sum of the squared errors. Thus, we take the errors shown in (2.1b), square them, and then take the sum over all observations in our sample (the sample size is indicated by n). Doing this, we have

$$\sum_{i=1}^{n} e_i^2 = \sum_{i=1}^{n} (Y_i - a - bX_i)^2.$$ (2.2)

At this point, our task now becomes a calculus problem. We have an equation (Equation 2.2) that we want to minimize with respect to two parameters we can choose, a and b. To carry out this task, we would use differential calculus and take the partial derivative of (2.2) with respect to a and set it equal to 0, then do the same with respect to b. We would then end up with two equations (the two partial derivatives) with two unknowns (a and b). The last step would then be to solve this system of equations simultaneously for a and b. Leaving out the details, what we end up with is the following formulas for b and a[3]:

$$b = \frac{\sum_{i=1}^{n}(X_i - \overline{X})(Y_i - \overline{Y})}{\sum_{i=1}^{n}(X_i - \overline{X})^2} \qquad (2.3a)$$

and

$$a = \overline{Y} - b\overline{X}, \qquad (2.3b)$$

where in both (2.3a) and (2.3b) the bar above the variable stands for the mean of that variable (e.g., the sum of all X_i divided by the sample size, n). Thus, given values for X_i and Y_i, we can use these data to calculate a and b.

As an illustration, we can go back to our baseball example. Using the data on salary (Y_i) and years of MLB experience (X_i) provided in Table A.1 in Appendix A, we can construct Table 2.1, which has the components needed for computing b and a.[4]

Viewing Table 2.1, we see that the sum at the bottom of the sixth column is the value needed for the numerator in (2.3a). The sum at the bottom of the fifth column is the valued needed for the denominator. Plugging these values into our equation for b (and rounding to three decimal places), we have the following:

$$b = \frac{228.836}{653.875} = 0.350. \qquad (2.4a)$$

As for the intercept term, a, dividing the sums at the bottom of the first and second columns by our sample size, $n = 32$, we find

$$\overline{X} = \frac{\sum_{i=1}^{n} X_i}{n} = \frac{242}{32} = 7.562,$$

$$\overline{Y} = \frac{\sum_{i=1}^{n} Y_i}{n} = \frac{149.701}{32} = 4.678.$$

TABLE 2.1

X_i	Y_i	$X_i - \overline{X}$	$Y_i - \overline{Y}$	$(X_i - \overline{X})^2$	$(X_i - \overline{X})(Y_i - \overline{Y})$
1	0.180	−6.563	−4.498	43.066	29.519
11	5.442	3.438	0.764	11.816	2.625
7	7.945	−0.563	3.267	0.316	−1.838
12	8.917	4.438	4.238	19.691	18.808
5	2.800	−2.563	−1.878	6.566	4.813
7	3.583	−0.563	−1.095	0.316	0.616
2	0.400	−5.563	−4.278	30.941	23.797
12	8.000	4.438	3.322	19.691	14.741
9	7.979	1.438	3.301	2.066	4.745
8	5.000	0.438	0.322	0.191	0.141
1	0.230	−6.563	−4.448	43.066	29.191
16	4.000	8.438	−0.678	71.191	−5.722
8	2.750	0.438	−1.928	0.191	−0.844
3	2.425	−4.563	−2.253	20.816	10.280
8	6.500	0.438	1.822	0.191	0.797
6	4.500	−1.563	−0.178	2.441	0.278
10	8.333	2.438	3.655	5.941	8.909
19	4.250	11.438	−0.428	130.816	−4.897
11	5.450	3.438	0.772	11.816	2.653
5	8.000	−2.563	3.322	6.566	−8.512
17	6.300	9.438	1.622	89.066	15.306
3	2.113	−4.563	−2.566	20.816	11.706
7	6.600	−0.563	1.922	0.316	−1.081
5	5.000	−2.563	0.322	6.566	−0.825
8	8.000	0.438	3.322	0.191	1.453
2	0.325	−5.563	−4.353	30.941	24.215
8	7.000	0.438	2.322	0.191	1.016
9	5.250	1.438	0.572	2.066	0.822
7	6.600	−0.563	1.922	0.316	−1.081
3	0.750	−4.563	−3.928	20.816	17.922
11	4.800	3.438	0.122	11.816	0.419
1	0.280	−6.563	−4.398	43.066	28.863
Sums: 242	149.701			653.875	228.836

Using these values for the mean of X and Y, along with the value for b found in (2.4a), we have for our intercept (again, rounding to three decimal places):

$$a = 4.678 - (0.350) * 7.562 = 2.031. \tag{2.4b}$$

Thus, putting it all together, the OLS regression line is

$$\widehat{Y}_i = 2.031 + 0.350X_i, \tag{2.5a}$$

where the "hat" above Y_i denotes the predicted salary for player i. As an alternative way of reporting results, we can replace Y and X with their variable names:

$$\widehat{SALARY}_i = 2.031 + 0.350(YEARS_i). \tag{2.5b}$$

Interpreting these results, the intercept term, 2.031, is the predicted salary (in millions of dollars) for a player with no MLB experience (i.e., a rookie).[5] The value of b, 0.350, represents the added salary that a player earns, on average, for each additional year he plays MLB. Or, an additional year of experience adds, on average, about \$350,000 to salary, all else being equal. Thus, a player who has 5 years of MLB experience is expected to earn

$$\widehat{SALARY}_i = 2.031 + 0.350(5) = 3.781. \tag{2.5c}$$

Comparing this predicted salary to MLB player Bret Boone, who has 5 years of experience (see Table A.1), we see that his actual salary for the 1998 season was \$2.8 million, which is about \$1 million *less* than what his predicted salary would be according to (2.5c). On the other hand, if we consider the salary for 5-year player Tim Salmon, his actual salary was \$5 million, or about \$1.2 million *more* than predicted. The difference between these actual figures and the predicted ones is captured by the error term, e_i, and demonstrates the imperfect relationship between salary and years of experience. Essentially, what we learn from this result is that years of experience may be important but there are other factors in addition to experience that determine a player's salary.[6] Figure 2.3 shows the regression line [(2.5a)] plotted with the actual data for player salaries and years of experience. The fact that the plotted points do not fall precisely on the regression line illustrates this last point.

Applying the same least-squares method to our example for U.S. presidential elections, we can estimate the relationship between economic growth and presidential voting patterns. Using the data shown in Table A.3, we could plug the values for the dependent variable (the column of data labeled "Votes") in for Y_t and the values for the independent variable (the column of data labeled "Growth") in for X_t

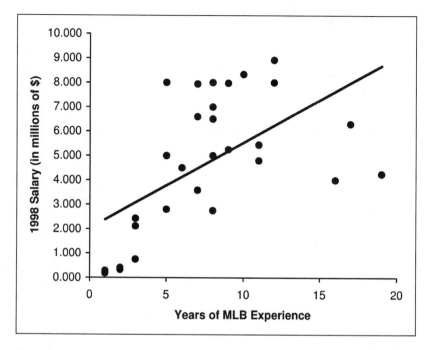

Figure 2.3.

into (2.3a) and (2.3b) to determine the value for the slope, b, and the intercept, a, for the sample regression function shown in (1.6). Although it was instructive to do this calculation by hand in the previous example on baseball, these calculations can be performed more easily by using computer programs designed to carry out these tasks. There are *many* such programs that are capable of calculating sample regression lines.[7] Perhaps two of the most prevalent and commonly used programs are Microsoft Excel and SPSS. To give the reader experience in reading and interpreting regression analysis output from these programs, the examples presented from this point forward will use these two programs (on an alternating basis) to carry out the calculations.[8]

Using Excel, we can input the data presented in Table A.3 for the column labeled "Votes" and the column labeled "Growth" into a spreadsheet. As shown at the bottom of Table A.3, the variable VOTES is defined to be the percentage of the two-party vote received by the incumbent-party candidate. This is our dependent variable Y_t in (1.6). The data for GROWTH is the growth rate of real gross domestic prod-

TABLE 2.2

	Summary Output	
Observations	21	
	Coefficients	
Intercept	51.5258118	
Growth	0.871151363	

RESIDUAL OUTPUT

Observation	Predicted Votes	Residuals
1	53.46760819	−1.787608186
2	41.53980373	−5.41980373
3	48.15271372	10.08728628
4	55.55314455	3.266855453
5	37.95850048	2.881499521
6	62.52409775	−0.06409775
7	53.6339981	1.366001904
8	54.06086226	−0.290862263
9	54.23073678	−1.860736779
10	52.31855954	−7.718559538
11	50.23737893	7.522621067
12	51.54323483	−1.633234826
13	55.83801104	5.501988957
14	55.63067702	−6.030677019
15	56.50531299	5.284687013
16	54.4973091	−5.547309096
17	48.46632821	−3.766328213
18	56.51053989	2.659460105
19	53.41969486	0.480305139
20	52.81337351	−6.263373512
21	53.26811452	1.331885476

uct (GDP) over the three quarters prior to the election.[9] This is our independent variable X_t shown in (1.6).

Once the data have been input, we can then follow the steps for calculating the least-squares regression values for a and b. Doing so yields the results shown in Table 2.2.[10] The first entry, "Observations," simply reports that we have 21 observations in our sample. Next, we see a column headed "Coefficients" and two rows labeled "Intercept" and "Growth." The entries are shown as (rounding to three decimal places) 51.526 and 0.871. These are the least-squares values for the intercept term *a* and the slope term *b* (respectively) for (1.6). Using these results, we can now write the predicted equation as

$$\widehat{Y}_t = 51.526 + 0.871X_t. \tag{2.6a}$$

Once again, we may reexpress our results using variable names in place of Y and X:

$$\widehat{\text{VOTES}}_t = 51.526 + 0.871(\text{GROWTH}_t). \tag{2.6b}$$

This equation tells us that if the growth rate were 0 over the three quarters prior to the election, then the incumbent-party candidate would be expected to receive approximately 51.526% of the two-party vote. In addition, for every 1% increase of real GDP over the last three quarters prior to the election, the incumbent-party candidate is expected to gain approximately 0.871% of the two-party vote. This, of course, works in the other direction. That is, for every 1% decline in the economy's real GDP, the incumbent-party's candidate is expected to suffer a 0.871% loss of the two-party vote.

The last part of Table 2.2 shows predicted values and the associated residuals for 21 elections covered in our sample of data. The column headed "Predicted Votes" shows the percentage of the two-party vote the incumbent-party candidate was predicted to receive according to (2.6b), plugging in the actual value of GROWTH. That is, for observation 1 (the 1916 presidential election), the growth rate reported for that year is 2.229 (from Table A.3). If we plug this value into (2.6b), we would have the following predicted value for that election:

$$\widehat{\text{VOTES}}_1 = 51.526 + 0.871(2.229) = 53.468. \tag{2.6c}$$

However, the actual value for that year was 51.68. The fact that our predicted value is not equal to the actual value again represents the fact that our simple model is not a perfect one. The difference between the actual values and the predicted values is what is shown in Table 2.2 as the "Residuals"—the "*e*" in our equation (1.6). Thus, for the first observation, we have (rounding to two decimal places)

$$e_1 = \text{VOTES}_1 - \widehat{\text{VOTES}}_1 = 51.68 - 53.47 = -1.79. \tag{2.7}$$

This result shows that our model overpredicted the actual percentage of the two-party vote received by the incumbent-party candidate

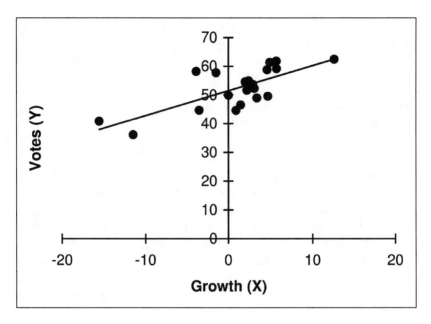

Figure 2.4.

(Woodrow Wilson, in this case) by 1.79%. The residuals are calculated for each election in our sample and are reported in Table 2.2.

The regression line shown in (2.6a), along with the actual values for Y_t and X_t, is plotted in Figure 2.4.[11] As shown in Figure 2.4, the actual values do not fall precisely on the regression line, but are speckled above and below it. Again, this illustrates the imperfect relationship between Y_t and X_t, with the vertical distance from any dot to the regression line equaling the error in prediction.

Turning now to our example of state abortion rates, we can use the same tool of least-squares to estimate the sample regression function shown in (1.7). Recall that, for this model, the dependent variable, Y_i, is the abortion rate for state i, and the independent variable, X_i, is the measure for the variable that was called RELIGION. In this case, we will use the program SPSS to calculate the least-squares values for a and b in (1.7). Doing so, we obtain the results given in Table 2.3.[12] We see in Table 2.3, under the "Unstandardized Coefficients" heading, that the intercept, or "constant" term, is 23.825. This is our value for a in (1.7). The value for b is the coefficient for RELIGION and is shown

TABLE 2.3

		Coefficients[a]			
		Unstandardized Coefficients			
Model		*B*	Std. Error	*t*	Sig.
1	(Constant)	23.825	3.979	5.988	0.000
	RELIGION	$-9.94E-02$	0.114	-0.874	0.386

[a]Dependent variable: ABORTION RATE.

to be $-9.94E-02$. This number is given in "scientific notation," which is a common way for output to be displayed in regression programs. The $E-02$ indicates that the number -9.94 is being multiplied by 10 to the power -2. Thus, b is -0.0994. Using these values for a and b, we have the following sample regression function (rounded to three decimal places):

$$\widehat{Y}_i = 23.825 - 0.099X_i. \qquad (2.8a)$$

Or, using variable labels,

$$\widehat{ABORTION}_i = 23.825 - 0.099(RELIGION_i). \qquad (2.8b)$$

This equation for the sample regression function is plotted in Figure 2.5[13] along with the actual data. The dots representing actual observations are not closely speckled around the regression line, indicating that although X_i (religion) may explain some of the behavior of Y_i (the abortion rate), there is a great deal that is left unexplained. As we have discussed, this unexplained portion is captured by the error term e_i and is equal to the vertical distance from the dots shown to the regression line. This example illustrates the point made in Chapter 1 that in many cases our simple bivariate model will not be sufficient for explaining the behavior of a dependent variable and a multivariate regression model will be needed. The subject of multivariate models will be taken up later in Chapter 4.

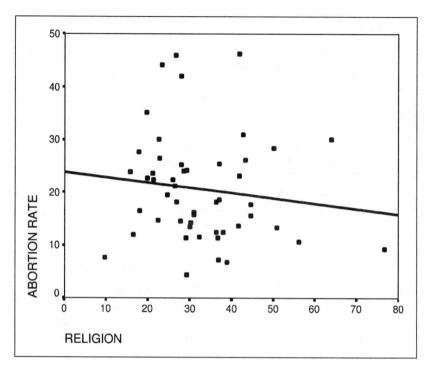

Figure 2.5.

Regression Model Assumptions and the Properties of Ordinary Least-Squares

The OLS method of estimating regression lines is clearly a powerful research tool. The validity of the OLS results we obtain, however, depends on a series of assumptions, called the classical linear regression model (CLRM) assumptions, which we have yet to discuss. These assumptions are sketched briefly in the following discussion.[14] The end result is that if these assumptions are satisfied, then the OLS estimated regression line gives us the best possible representation of the population's regression line.[15]

Classical Linear Regression Model Assumptions

1. The average of the population errors (u_i's) is 0. As shown in Figure 1.2(b), some points will lie above the population regression

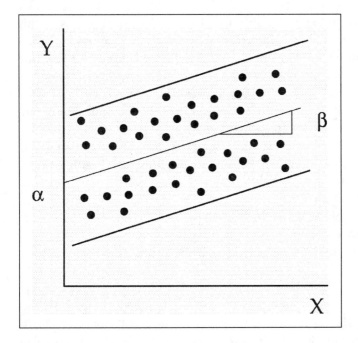

Figure 2.6.

function and will have positive errors, and some will lie below and have negative errors. On average, the errors should cancel each other and thus the average of the errors should be 0.

2. The spread of the errors above and below the regression line (i.e., the **variance**) is uniform for all values of X. Graphically, this means that the actual observations for Y_i, for given values of X_i, fall within a uniform band around the population regression function, as shown in Figure 2.6. As shown, the population regression function (PRF) has observations that are above and below it, but they are uniformly spread around the line, as the two darker, parallel lines above and below the PRF demonstrate. (The technical term is that the errors are said to be **homoscedastic**, meaning "equal variance." An example of when the assumption is violated is given later in Chapter 6.)

3. The error associated with one observation is not associated with errors from any other observations. Or, in technical terms, we as-

sume no **autocorrelation** among the error terms. The basis for this assumption is straightforward. The errors are supposed to represent purely random effects, which our model is unable to control. If, however, one observation's error is somehow related to another observation's error, then this implies that there is some systematic relationship among the errors and thus they are not purely random. The implication is that this systematic relationship contains information that we should use to improve the estimation of our model. If it is ignored, then we are not fitting the best line to our data. (An example of this kind of problem is considered in Chapter 6.)

4. The variables X_i and Y_i contain no measurement errors. Again, the necessity of this assumption is easy to understand. If X_i and Y_i are measured inaccurately, then the OLS values for a and b [which, as (2.3a) and (2.3b) show, are derived from the values of X_i and Y_i,] are not likely to be accurate estimates of the population's α and β.

5. Our fifth assumption is that the regression model is specified correctly. That is, the model we put forth, such as the one in (1.2b), is theoretically sound. This assumption can be violated in a number of ways, including omitting relevant independent variables. For example, if we consider our baseball salary model, we hypothesized that years of playing determine a player's salary. We know, however, that years alone do not determine baseball salaries. A player's offensive ability (e.g., hitting) or defensive ability (e.g., fielding) is also obviously important. If we do not take into account these important factors in our model, then our model will not be correctly specified (i.e., we commit a **model specification error**), and the values for a and b may not be reliable.

   Another way we can misspecify our model is to use a functional form that is inappropriate. For example, we may fit a straight line to data when, in fact, a curve is appropriate. (We will have more to say about functional forms later in Chapter 6.)

6. Our last assumption has to do with how the population's error term, u_i, is distributed. We assume that the u_i follows a normal distribution (often referred to as the **normality assumption**). That is, the random errors follow the familiar bell-shaped curve that is well known from statistics. The importance of this assumption will be seen later in Chapter 3.[16]

If CLRM assumptions 1 through 5 are satisfied, then, as noted previously, the OLS regression line provides the *best possible* estimate of the population regression line. Or, using the more common terminology, OLS is **BLUE**, which is an acronym for *best linear unbiased estimator.* To understand this property, we can begin by discussing what is meant by **linear** and **unbiased**. Linear simply means that we are estimating values for the intercept (a) and the slope (b) that are raised only to the power 1. Thus, (1.5) is an example of a linear regression model. However, consider the following equation:

$$Y_i = a + b^2 X_i + e_i. \qquad (2.9)$$

In this case, b is raised to the power 2 and is thus not "linear" in the sense described here. On the other hand, we do allow for the independent variables to enter into our equation nonlinearly as shown in the following equation:

$$Y_i = a + bX_i^2 + e_i. \qquad (2.10)$$

This is considered a linear estimation because a and b are raised to the power 1 and thus (2.10) is a candidate for the BLUE property of OLS.[17]

Unbiasedness has to do with the fact that our sample estimate of b, the slope term, is a random variable. That is, given that we use a sample to calculate b, if we repeat the estimation with new samples we will likely find different values for b. If we do so, then we can calculate *the average of all these b's.* If it is true that the average of the b's is equal to the population's true β, then the estimator is said to be unbiased. Similarly for a, the intercept term.[18]

Now that we know what is meant by linear and unbiased, we can explain what is meant by "best." If we consider all possible estimation methods that are linear and produce unbiased estimates of a and b, the OLS method is the best one in the sense that it gives us the most precise estimates of a and b. To understand the meaning of this last statement, recall that the estimated parameters a and b "bounce" around from sample to sample (i.e., they are "random variables"). If they are unbiased, this means that as they bounce around they have a mean value that is equal to the population's α and β. To be best, it will be the case that they bounce around the least; any other linear unbiased estimation method will produce values of a and b that bounce around more than those calculated using the OLS method. In other words, of all linear unbiased estimation methods, the OLS method gives us the most precise estimates of α and β, or OLS is BLUE.[19]

Thus, we have powerful support for the use of OLS.

SUMMING UP

In this chapter, we have seen how we may use samples of data and employ the least-squares method to estimate the linear relationship between a dependent variable and an independent variable. We have seen, however, that our estimated sample regression function does not completely explain the relationship between the dependent variable and the independent variable and what is left unexplained shows up in the error term. The question we now turn to is that of model performance and reliability. That is, once we have estimated a relationship between a dependent variable and an independent variable, what can we say about *how well* we have estimated this relationship? This is the topic of the next chapter.

▼ Notes

1. We could consider minimizing the sum of the absolute value of the errors, but such a method is computationally difficult.

2. There are other, advanced methods such as "two-stage least-squares" or "weighted least-squares" that can be used in certain circumstances. These methods are beyond the scope of this book. See, for example, Gujarati (1995) for a discussion of these techniques and others.

3. For the full details of solving of the ordinary least-squares (OLS) estimators, see Gujarati (1995).

4. Our analysis will consider only nonpitchers because pitchers are evaluated with very different statistical measures.

5. The average rookie salary is, in fact, much lower. The result from this regression is higher than the MLB average rookie salary because the sample used (which was hand picked) does not closely resemble MLB as a whole. This should serve as a word of caution to the reader that *sample* results may not always be a good reflection of the population's behavior.

6. Explaining part of the discrepancy between Boone and Salmon's salary is the fact that Salmon has a better career slugging average of 527 (through the 1997 season) compared to Boone's career slugging average of 327.

7. For example, some popular programs include SAS, TSP, RATS, MINITAB, SHAZAM, STATA, etc.

8. Microsoft Excel is a relatively easy program to use and performs the basic calculations we will need for this book. SPSS is specifically designed to do statistical calculations and is capable of carrying out more sophisticated analyses. Appendix B provides some basic instruction on how to perform regression analysis using Excel and SPSS. The reader is referred to the instruction manuals and tutorials that accompany these programs for greater details on how to use these programs. Additionally, Einspruch (1998) gives detailed instruction on how to use SPSS.

9. The U.S. GDP is a measure used by economists to track the growth of the U.S. economy. It is defined as the total market value of all final goods and services produced inside the United States over a specified period of time.

10. The results presented in Table 2.2 are an edited version of the actual output Excel produces. Much of the actual output the program produces was excluded in this example to make things easier for the reader. The full output will be presented in later examples as we progress.

11. This graph was generated by the Excel program.

12. As in the previous example using Excel, the actual output produced by SPSS was edited so as to include only the results relevant to the discussion at hand. The full SPSS output will be presented in later chapters.

13. Figure 2.5 was produced by the program SPSS.

14. For a more detailed discussion of the regression model assumptions, see Berry (1993).

15. A formal proof that the OLS method is the best one is given by the famous Gauss–Markov theorem. For more details, see Greene (2000).

16. The justification of this assumption comes from a theory from statistics called the "central limit theorem." Those interested in learning more about this theorem are referred to Greene (2000).

17. Our linearity requirement is only for the parameters to be estimated (i.e., a and b). The use of nonlinear independent variables, such as shown in (2.10), are discussed later in Chapter 5.

18. Formally, if it is true that

$$E(a) = \alpha \quad \text{and} \quad E(b) = \beta,$$

where E stands for expected value, then we say that a and b are unbiased estimators of α and β. It then follows that if $E(b) \neq \beta$, then b would be a biased estimator of β. Similarly for the intercept term, a.

19. In technical terms, OLS estimates of the random variables a and b will have the smallest variance as compared to any other estimation method. For a more detailed discussion, see Gujarati (1995) or Berry (1993).

PROBLEMS

2.1 Consider the following data set:

Y_i	X_i
10.5	13
9.75	12
10.00	12
12.25	14
8.00	10

where Y_i is individual i's hourly wage (in dollars per hour) and X_i is individual i's number of years of education. Use (2.3a) and (2.3b) to calculate the OLS values for a and b and interpret your results.

2.2 Use (2.5b) and the data in Table A.1 of Appendix A to predict the salary of Mark McGwire. What is the error in prediction (i.e., e_i)? What may account for this error?

2.3 Suppose we have the following model:

$$Y_i = \alpha + \beta X_i + u_i,$$

where, in this case, Y_i is the manufacturer's suggested retail price (MSRP) for a sports utility vehicle (SUV) and X_i is the horsepower of the SUV.

a. What sign do you expect for α and β?

b. Using SPSS or Excel (or an equivalent program) and the data provided in Table A.5, perform an OLS regression for the preceding model and interpret the estimated coefficients.

MODEL PERFORMANCE AND EVALUATION

In the last chapter, we learned how to use the powerful method of least-squares to find a line that best fits a scatter of points. It can be shown that, given the right circumstances, there is no other method that is better at estimating such a line than least-squares. That is, if the CLRM assumptions are met, then OLS is BLUE. At this point, two other issues arise. First, suppose we have used least-squares to find a line that best fits the data. We can then ask *how well* does the line fit the data. Finding the best fit is one thing, but how well our line fits the data is quite another. This is the issue of "goodness of fit," which we explore in the next section.

Another question we can ask is not how well the regression model as a whole performs, but how the separate pieces of the model perform. That is, in the previous chapter we learned how to calculate the least-squares values of *a* and *b* to determine the sample regression line. Because these values were derived from a sample, however, they may not be representative of the *population's* α and β. Thus, we can ask the question: How confident are we that our sample results are a good reflection of the population's behavior? We discuss this issue later in the chapter.

Goodness of Fit: The R^2

In Chapter 2, we estimated three sample regression lines, one for each of our examples. We then plotted these lines along with the actual data for the dependent variable, *Y*, and the independent variable, *X*. We have already accepted the fact that in each case the dots representing the data will not fall exactly on the regression line, reflecting the fact that our model does not take into account all factors that affect our dependent variable. In general, however, we hope that the vertical distances from the dots to the regression line are small because, if this is true, then our estimated line would be a good predictor of the behavior of Y.

Reviewing Figures 2.3, 2.4, and 2.5, we can see some qualitative differences. In Figure 2.3, which is for our baseball salary example, we see that the plotted data points are somewhat broadly scattered around the sample regression line. This indicates that the model explains some of the behavior of Y, but much is left unexplained. Comparing this to Figure 2.4, which is for the presidential voting model, we see that the plotted points in this case are more closely cropped around the regression line. Thus, for this example, the regression line seems to tell us a lot about the behavior of the dependent variable, Y. Finally, in Figure 2.5, we see that the plotted data for state abortion rates and religion are spread very broadly around the regression line shown there. In this case, we can conclude that although religion may be an important factor in determining abortion rates, there are many other important factors we need to consider.

Even though the least-squares regression method produces the best possible line to fit our data[1] all this means is that we have done the best we can and the overall performance of the model is not guaranteed to be good. To judge how well the model fits the data, we can employ a measure called the R^2 (read as the **R squared**).[2] The technical derivation of this measure can be a little hard to follow, but the intuition of it is not too difficult to understand.

Our basic approach as set out in Chapter 1 is to understand the behavior of Y, the dependent variable, by observing the behavior of X, the independent variable. As the value of X differs from one observation to another, we expect that this difference in X will explain, at least in part, the differences in Y from one observation to another. This approach is summarized by (1.5). We hope that by observing the behavior of X we can learn a great deal about the behavior of Y. That behavior of Y that is not explained by X will be captured by the error term in (1.5). *The R^2 simply is a measure that tells us what percentage of the behavior of Y is explained by X.*

We can easily consider the bounds for the R^2. If, by observing the behavior of X, we know exactly how Y behaves, then the R^2 would be equal to 1 (100%). This outcome is very unlikely.[3] On the other hand, observing X may tell us nothing about Y, in which case the R^2 would be equal to 0 (0%). Thus, the R^2 is bounded between 0 and 1. Values close to 1 mean that by observing the behavior of X we can explain nearly all of the behavior of Y. This would indicate that our estimated sample regression function is performing well, overall. Values close to 0 would have the opposite implication.

The preceding discussion gives us an intuitive understanding of the R^2. To understand the technical derivation of this measure, we need to define what is meant by the "behavior" of Y. The dependent variable Y will have a mean (average) and, of course, some values of Y will fall above this mean and some below. As a graphic example, consider Figure 3.1. As shown in Figure 3.1, a particular point is plotted for observation Y_i for a given value of X_i. Of course, there would normally be many other observations plotted on the graph, but they are not shown in this case so that we may focus on this single observation. The mean

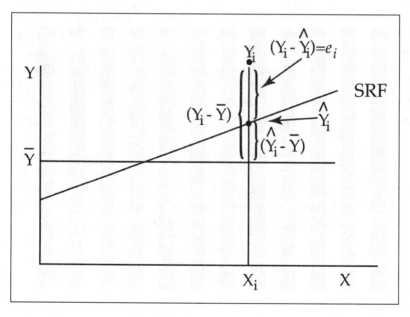

Figure 3.1.

value for the dependent variable, $\overline{Y}$, is a horizontal line. The sample regression function (SRF) is also plotted. Notice that the plotted point Y_i lies above $\overline{Y}$ (i.e., it is above the horizontal line). Thus, the value of Y for this particular observation is above the sample's average value for Y. The amount by which Y_i exceeds $\overline{Y}$ is shown as the distance from the horizontal line to the point Y_i and is denoted as $(Y_i - \overline{Y})$. This deviation of Y_i from its mean can be broken up into two pieces. That part of the deviation that our model predicts, shown as the vertical distance from the mean of Y to the value the model would predict for Y given X_i (i.e., the point on the sample regression function above X_i), is denoted as $(\widehat{Y}_i - \overline{Y})$. And that part of the total deviation that was not predicted by our model, shown as the vertical distance from the point on the sample regression function to the observation Y_i, is denoted as $(Y_i - \widehat{Y}_i)$, which is simply our error term, e_i. Thus, suppose this observation is from our baseball example. Then the salary for player i, Y_i, is above the average salary for all players in our sample. Part of the amount by which player i's salary exceeds the average is explained by our model. That is, it can be explained by the number of years player i has played in the major leagues. The rest is not explained by our model and is captured by the residual, e_i. It should be obvious that the greater the proportion of the deviation of the observation from its mean value that is explained by our model, the better our model is performing because this would mean that the proportion accounted for by the error term is smaller. If, on the other hand, our model does a poor job in explaining the deviation of Y_i from its mean, then our error term will be larger. This kind of breakdown of the total deviation of Y_i from its mean into the explained portion and the unexplained portion can be done for all observations in our sample.

Now, as noted before, the R^2 is a measure that tells us what percentage of the behavior of Y is explained by X. We can now use the method described previously to give a more precise meaning to the "behavior of Y." We can define this behavior of Y as the *variation* in Y, which is calculated according to the following equation:

$$\text{TSS} = \sum_{i=1}^{n}(Y_i - \overline{Y})^2. \qquad (3.1)$$

That is, the variation in Y is the sum of the squared deviations of Y around its mean. We will call this sum the total sum of squares, or TSS for short. Essentially, it is a measure that tells us by how much the values of Y "bounce" around its mean. The deviations are squared so as to prevent cancellation of positive values with negative ones (recall this was also done when we derived the OLS method in Chapter 2).

Part of this behavior of Y (i.e., the TSS) is explained or "predicted" by our model; the rest is left unexplained. The part of this variation that is explained is

$$\text{ESS} = \sum_{i=1}^{n} (\widehat{Y}_i - \overline{Y})^2. \tag{3.2}$$

We call this value the explained sum of squares, or ESS for short. It represents the explained behavior of Y about its mean.

Finally, as we have noted, the error represents what is not explained by our model:

$$\text{RSS} = \sum_{i=1}^{n} e_i^2. \tag{3.3}$$

We call this sum the residual sum of squares, or RSS. Using these measures, we can then represent the "behavior" of Y in the following way:

$$\text{TSS} = \text{ESS} + \text{RSS}. \tag{3.4}$$

That is, the behavior of Y (TSS) can be broken up into two pieces: that which is explained by the model (ESS) and that which is the unexplained (RSS). The R^2 is thus defined as the proportion of the TSS explained by the model, or

$$R^2 = \frac{\text{ESS}}{\text{TSS}}. \tag{3.5a}$$

As an illustration of how to use the R^2, we can return to our second example, which looks at presidential voting. Looking at Table 2.2, we see the reported predicted values (i.e., the $\widehat{Y}_i$ values). In Table A.3 of Appendix A, we have the actual values for the percentage of two-party votes (i.e., the Y_i values) as well as the mean (i.e., the value for $\overline{Y}$). Thus, we have all the ingredients needed for calculating the R^2. Plugging these values into (3.1) and (3.2), we find that the TSS is 1014.248 and the ESS is 548.865. Using (3.5a) then, we have

$$R^2 = \frac{548.865}{1014.248} = 0.541155. \tag{3.5b}$$

The result in (3.5b) thus means that approximately 54% of the variation in Y is explained by our model. This, in turn, tells us that about 46% of the behavior of Y is not explained by our model. These results suggest that our model is moderately successful, overall, in explaining the variation in the percentage of two-party votes received by incumbent-party candidates.[4]

Virtually all computer packages that are capable of performing OLS regression will calculate the R^2. For example, Table 3.1 shows the OLS output generated by Excel for our baseball model. As we can see, Table 3.1 provides a great deal of information.[5] We can see under the heading "Coefficients" the values for our intercept term (2.031) and the slope term, b (0.350), shown earlier in (3.5a), after rounding to three decimal places. At the top of the table, we see the heading "Regression Statistics" below which we find "R Square," which is reported as approximately 0.3347. This is our measure of the "goodness of fit" discussed previously. The interpretation of this number is that about 33.47% of the behavior (i.e., variation) of baseball salaries is explained by our model. This means that about two thirds of the variation in baseball salaries is left unexplained. This somewhat poor result should not

TABLE 3.1

Summary Output

Regression Statistics

Multiple R	0.578533258
R Square	0.334700731
Adjusted R Square	0.312524088
Standard Error	2.30354598
Observations	32

ANOVA

	df	SS	MS
Regression	1	80.08563793	80.08563793
Residual	30	159.1897224	5.306324081
Total	31	239.2753604	

	Coefficients	Standard Error	t Stat	P Value
Intercept	2.031527236	0.793688934	2.55960131	0.015759687
YEARS	0.349969303	0.090084385	3.884905278	0.000523394

be too surprising because, as was pointed out in Chapter 1, the two-variable model is too simplistic in most cases and a more complex model is warranted.

Also shown in Table 3.1 is a section labeled "ANOVA," which stands for *analysis of variance*. Focusing on this section for a moment, we see a column headed "SS." This is short for sum of squares and is a breakdown of the variation of Y (baseball salaries) into its separate pieces as described in (3.4).[6] The first entry for "Regression," shown as approximately 80.086, is our ESS in (3.4). The second, labeled "Residual," shown as approximately 159.120, is our RSS in (3.4). Finally, "Total," given as 239.275 after rounding to three decimal places, is our TSS (= ESS + RSS) discussed earlier. Thus, using our definition of the R^2 in (3.5a), we have

$$R^2 = \frac{80.086}{239.275} = 0.3347,$$
(3.5c)

which is indeed our reported "R Square" in Table 3.1.[7]

There are several other measures shown in Table 3.1 under the heading of "Regression Statistics" that we can discuss briefly. The "**Standard Error**," shown as approximately 2.303, is the positive square root of the variance of the errors. It is essentially a measure of the typical size of the error (our e_i) in prediction. "Observations" simply tells us the sample size (32 in this case) used in the regression.

Sample Results and Population Parameters

In addition to judging the overall performance of our model, we can consider the separate performance of the estimated parameters, a and b. It was pointed out in Chapter 2 that we typically are working with samples of data, not the population. Thus, we collect a sample and use it to calculate OLS values for a and b; these values then define a sample regression line. The hope is that our sample values of a and b are a good representation of the true, but unknown, values of the population parameters α and β shown in (1.2b). Suppose we replaced our original sample and collected a new one. This new sample could then be used to calculate OLS values for a and b. Obviously, because this second sample would not likely be identical to the previous one, these values for a and b would likely be different from those of the first sample. Different samples have different information contained in them.

Thus, as samples change so will the OLS-calculated values for a and b. In other words, these sample values for a and b are random variables that "bounce around" from sample to sample and we can thus study their behavior. To illustrate the fact that a and b typically change from sample to sample, we can return to our baseball example. Suppose we draw two samples of size 16 from our 32 observations. We can then calculate the OLS values for a and b for each sample. We have for observations 1 through 16:

$$a = 0.921, \qquad b = 0.482$$

and for observations 17 through 32 we have

$$a = 2.959, \qquad b = 0.252.$$

As we can see, we have substantially different values for these estimated parameters from one sample to the next.

That the OLS-calculated values for a and b are, in fact, random variables is important because we use them as estimates of the population's α and β. If the calculated values of a and b change a lot from sample to sample, then we would not have much confidence in any single sample's results being representative of α and β. On the other hand, if the OLS estimates of a and b differ only slightly from one sample to another, then we can be fairly confident that a single sample's results are a good representation of the population. The measures used to judge the reliability of a and b as estimates of their population counterparts are the standard error of the coefficient estimates.[8] Essentially, they are measures of how a and b bounce around from sample to sample. The *smaller* the standard error, the *more reliable* are the sample values of a and b as estimates for α and β. In our baseball example, we can see in Table 3.1 that the column headed "Standard Error" for the "Intercept" (i.e., a) is shown as 0.794 and for "YEARS" (i.e., b) we have 0.090, after rounding to three decimal places. There are two things that we need to keep in mind with these standard errors: First, they are in the same units as the dependent variable (dollars in our case). Second, it is their size *relative* to the value of the estimated coefficient that is important for us. That is, consider the value for b, which is shown in Table 3.1 to be 0.350. Comparing this number to its standard error, we see that the coefficient is nearly 4 times as large as the standard error. Thus, even though this standard error is a large number, it is small relative to the value estimated for b. In other words, the sample value for b appears to be a fairly reliable estimate of the population's β.[9]

Given the values of the standard errors of our a and b, we can use them to test various ideas or "hypotheses." One of the most common tests is the "zero hypothesis" test. This test, in some sense, looks into the "soundness" of our model. Recall Chapter 1 where we introduced the three examples. In each case, we specified a dependent (Y) variable and an independent (X) variable that we believed (or "hypothesized") should explain at least part of the behavior of the dependent variable. Take, for example, our model of voting patterns in presidential elections. In building this model, we hypothesized for the population data that the percentage of two-party votes received by the incumbent-party candidate is determined, in part, by the economy's health as measured by the growth rate. This relationship between votes and growth is not an established fact, but rather it is a theory or hypothesis that we put forth. It could be wrong; there may be no relationship between votes and growth. Part of our task in regression analysis is to test this hypothesis. We do this by collecting a sample, estimating the sample regression function, and then analyzing our results to see whether the hypothesis is supported or not. More specifically, consider our model shown in (1.6), except now for the population of data: $Y_t = \alpha + \beta X_t + u_t$. If economic growth (i.e., variable X_t) is an important factor in explaining voting patterns (i.e., Y_t), then the population's value for β should be positive. On the other hand, suppose that the economic growth rate has nothing to do with the way people vote. Then, in this case, β would be 0. If β is truly 0, then βX_t would be 0 meaning that X_t has no impact on Y_t.

Whether or not β is 0 is a hypothesis made for our population of data. Stating the hypothesis a bit more formally, we can write

$$H_0: \beta = 0,$$
$$H_1: \beta \neq 0,$$

where H_0 is called the **null hypothesis** and H_1 is the **alternative hypothesis**. If we find that H_0 is true, then our hypothesis that economic growth is an important factor in explaining voting patterns would be false. On the other hand, if we find that H_1 is true, then our hypothesis is correct. Unfortunately, we typically cannot *prove* that either H_0 or H_1

TABLE 3.2

	Summary Output		
Regression Statistics			
Multiple R	0.735632156		
R Square	0.541154669		
Adjusted R Square	0.517004915		
Standard Error	4.949124961		
Observations	21		
ANOVA			
	df	SS	MS
Regression	1	548.8649945	548.8649945
Residual	19	465.3829198	24.49383788
Total	20	1014.247914	
	Coefficients	Standard Error	t Stat	P Value
Intercept	51.5258118	1.098803993	46.89263245	4.17077E−21
Growth	0.871151363	0.184030274	4.733739416	0.000144496

is true. This is because we typically do not have the population data and are working only with a sample. Because our sample result for b is not a perfect predictor for β, our sample results can only *support* or *fail to support* our hypotheses over β.[10]

To see how we can use our sample regression results to test the previous hypotheses, we can use the information in Table 3.2, which contains the OLS output (produced by Excel) for our voting model.[11] We see in Table 3.2 that the estimated values for a and b are as we saw earlier, 51.526 and 0.871, respectively (after rounding to three decimal places). We can also see that the standard errors for these values of a and b are approximately 1.099 and 0.184, respectively. If we next consider the column headed "t Stat" (short for t **statistic**), we see the rounded value of 46.893 for the intercept, a, and 4.734 for the coefficient to "Growth," or b. The basis of the t statistic actually stems from the normality assumption that was introduced in Chapter 2 in CLRM assumption 6. Although the discussion of the use of the t statistic is somewhat advanced, it boils down to this: If the population errors, u_i, are normally distributed, then it can be shown that the sample estimates for a and b follow a t distribution.[12] Given this result, we can use the t distribution to test hypotheses over the population parameters α and β with our sample estimates. Or, in other words, the normality assumption introduced in Chapter 2 now justifies our use of the t statistic here.

As for the value for each t statistic shown, they are calculated by dividing a coefficient by its associated standard error. That is, for the intercept, if we divide 51.526 by 1.099 we get 46.88, which is the t-statistic value shown (allowing for rounding differences). Similarly for the coefficient to "Growth" (i.e., our b), we have 0.871, which, if divided by its standard error of 0.184, gives us 4.734, the reported t statistic for "Growth." As noted earlier, the size of the standard error relative to the estimated coefficient is what is important to us. This is exactly what the reported t statistic shows: the coefficient divided by its own standard error. In the case of the intercept, our t statistic of 46.89 tells us that our calculated value of a is about 47 times larger than its standard error. In other words, the OLS value for a is a reliable estimate of the population's α as it bounces around very little from sample to sample. As for b, this coefficient is approximately 4.7 times larger than its own standard error, indicating that this sample value for b is a very reliable estimate for the population's parameter β. Exactly how reliable are these estimates? The answer to this question is our next task.

The column next to the "t Stat" column in Table 3.2 is headed "**P Value**." The numbers shown here for a and b are calculated using the t statistic.[13] These numbers represent *probabilities*, based on the associated t statistic, that we can use to test the zero hypotheses discussed earlier. The interpretation of these numbers is a little tricky. They represent what the probability would be of finding the values of a and b shown, if, in fact, *for the population* the null hypothesis were true. In other words, it sets up a straw man, that we hope we can knock down, and this straw man says that, for the population, X has no relationship to Y.

To make this clear, let's focus on the value for b shown in Table 3.2. This coefficient to "Growth" is estimated to be 0.871, based on our sample OLS results. This number, as we have discussed, is only an estimate of the population's true value for β. Furthermore, we have learned that this result for b will likely differ from sample to sample (i.e., it is a random variable). This being the case, the value of 0.871 could be quite different from β. We can therefore consider the following question: Could it be the case that the true value for β is, in fact, 0, and that our value of 0.871 is merely a result of our sampling? This is the question that the

P value addresses: What is the likelihood of getting a sample value of 0.871 when, in fact, it is true that β is 0? In our case, the P value for b is shown as approximately 0.000144. That is, the probability of getting the value of 0.871 for b when it is true that β is 0 is 0.000144 (or 0.014%). This is quite small, indicating that the hypothesis that β is 0 is probably not true. In other words, we would likely reject H_0: $\beta = 0$, in favor of H_1: $\beta \neq 0$.

We can also perform a similar test for the intercept term, a. That is, we can consider the hypothesis that the population's intercept is really equal to 0. Putting it formally, we would have

$$H_0: \alpha = 0,$$
$$H_1: \alpha \neq 0.$$

To test this hypothesis, we need only consider the P value for a, which is given as approximately $4.171E - 21$. This is in scientific notation, which means we move the decimal 21 places to the left. Obviously, this is an extremely small number. In terms of our hypothesis test, this means that the probability of obtaining a value for a of 51.526 from our sample when, in fact, the value for α is truly 0 is $4.171E - 21$ (or $4.17E - 19\%$). In other words, we can be quite confident that the true population's intercept α is not 0.

In the preceding example, we found very small P values, which indicated that we can reject H_0 for both the intercept and the slope term with great confidence. Suppose, however, we were not able to be so confident. That is, suppose the P value for b turned out to be 0.20. This would mean that the chance of getting the value of b we found for our sample, when the population's β is, in fact, 0, is about 20%. Would we be sufficiently confident that β is not 0 with this result? In other words, how large of a P value are we willing to accept before concluding that we cannot reject H_0? If the P value is 0.20 and we then declare that we reject H_0, then this means we have a 20% chance that our declaration is wrong.[14] How prudent we are in rejecting the null hypothesis (i.e., accepting the estimated relationship) may vary with circumstances. For example, in a model exploring the usage of a particular drug and the possibility of its use causing birth defects, we would want to be quite sure that a real relationship exists. It is standard practice to reject the zero hypothesis when the associated P values are 0.05 or smaller (in some cases, P values of 0.10 or smaller are used).

To solidify the concept of hypothesis testing, we can apply these tools to our third example. Recall that this model, which is shown in (1.7), hypothesizes that abortion rates across the 50 states differ, in part, because the moral views on abortion differ across the states. We use the variable RELIGION, which shows the percentage of the population that is Catholic, Southern Baptist, Evangelist, or Mormon, to capture the moral differences. As RELIGION (X_i) increases from one state to another, we expect that the abortion rate (Y_i) would be smaller, other things being equal. As in the previous example, this is a hypothesis that we put forth. It could be wrong, meaning that RELIGION tells us nothing about abortion rates. That is, it may be the case that β is 0 for the population. We can test this hypothesis by evaluating the P value from the OLS regression. Table 3.3 shows the output from our earlier regression, but with more details in this case.

The three panels of output shown in Table 3.3 were produced by the program SPSS. The output provided is similar to that produced by the Excel program. We can see in the first panel, entitled "Model Summary," that the R^2 is reported as 0.016. This result tells us that approximately 1.6% of the behavior of abortion rates (Y_t) is explained by our model. This is clearly a very poor result, as it means that approximately 98.4% of the behavior (i.e., variation) of our dependent variable is not explained. This result is mirrored in the second panel, entitled "ANOVA," which shows that the residual sum of squares, RSS (reported as 4879.935), makes up nearly all of the total sum of squares, TSS (reported as 4957.626). In sum, the model, as it stands, fits the data very poorly.

Turning to the separate components of the model, we find similarly weak results. In the panel entitled "Coefficients," we find the estimate of a to be 23.825 with a standard error of 3.979, and the estimate of b to be −0.0994 with a standard error of 0.114. In the case of the constant term a, dividing the coefficient by its standard error (i.e., 23.825/3.979) gives us a reported "t" (which is the same as Excel's "t Stat") of 5.988. This t then translates into a "Sig." (short for significance level, which is the equivalent of Excel's P value) of 0.000. This result tells us that we can reject the hypothesis that the constant term for the population, α, is truly 0 with a very high degree of confidence.[15]

As for the coefficient to RELIGION (i.e., b), the results are quite different. The coefficient has the negative sign that was hypothesized, supporting the idea that as RELIGION increases from state to state,

TABLE 3.3

| Model Summary | | | | |

Model	R	R Square	Adjusted R Square	Std. Error of the Estimate
1	0.125[a]	0.016	−0.005	10.0829

[a]Predictors: (Constant), RELIGION.

| ANOVA[b] | | | | |

Model		Sum of Squares	df	Mean Square
1	Regression	77.691	1	77.691
	Residual	4879.935	48	101.665
	Total	4957.626	49	

[b]Dependent variable: ABORTION RATE.

| Coefficients[c] | | | | | |

Model		Unstandardized Coefficients		t	Sig.
		B	Std. Error		
1	(Constant)	23.825	3.979	5.988	0.000
	RELIGION	−9.94E−02	0.114	−0.874	0.386

[c]Dependent variable: ABORTION RATE.

abortion rates tend to be smaller, other things being equal. However, the value for b is not very different from 0. Again, we can ask the question: Is it true that the population value for β is really 0 and that our result for b of −0.0994 is from a sample that does not perfectly reflect the population? In other words, is −0.0994 *statistically* different from 0? The fact that the magnitude of our b is small is not enough to draw any conclusions about whether it is not statistically different from 0. We must compare the value for b to its standard error. The reported standard error of b is *larger* than the (absolute value) of b. In other words, for this model the random variable b tends to "bounce around" a great deal and as such the computed value is not a very reliable predictor of β. We can see this clearly by observing the "Sig." value for b, shown as 0.386. What this number means, then, is that there is a 38.6% chance of finding a sample value of -0.0994 for b when, in fact, the population's β is truly 0.[16] In other words, we cannot confidently rule out the possibility that RELIGION, on its own, has nothing to say about state abortion rates.

There are some occasions when we must work directly with the t statistic in order to test hypotheses. This may occur, for example, if the software program we are working with does not produce a P value (or significance level) for estimated coefficients.[17] Alternatively, we may wish to test a hypothesis other than the "zero hypothesis" test we have been performing. In both cases, we need to do some work by hand to perform a hypothesis test for our estimated coefficients. To see how we can work directly with the t statistic, we can return to our baseball example. As we saw earlier, the calculated value for the intercept is shown in Table 3.1 as 2.031 with a standard error of 0.794. Suppose we wanted to perform a zero hypothesis test for the estimated intercept, a, but we did not have a P value to use. To do so, we can use the t statistic provided and we can compare it to a table of t values that are provided in Appendix B. Before describing how to use the table of t values, however, we can explore the intuition behind this procedure.

Recall that the value of a is a sample's estimate of the population's intercept, α. The standard error of the intercept is a measure of how precise our estimate is: The larger the standard error, the less precise is our estimate. We noted that the "t Stat" reported by Excel (or the "t" reported by SPSS) is calculated by dividing the estimated coefficient by its standard error. That is, for the intercept, we have

$$t \text{ statistic} = \frac{a}{\text{se}(a)}, \tag{3.6a}$$

where se is short for standard error. Focusing on (3.6a) for the moment, we see that for any given value of a, the larger the se(a), the smaller would the t statistic. Therefore, the smaller the t statistic, the less reliable our value of a is as an estimate of the *population's* intercept, α. In other words, a "small" absolute value of the t statistic means that our value of a bounces around a lot from sample to sample. This, in turn, means that the population's value for α may truly be 0 and that our sample estimate of the intercept was simply off the mark. Putting this formally, we can state the following two hypotheses:

$$H_0: \alpha = 0,$$
$$H_1: \alpha \neq 0.$$

In this case, the "smaller" the absolute value of the t statistic, the more likely we would not reject H_0 (i.e., that the population's regression line has a zero intercept). On the other hand, the "larger" the absolute value of the t statistic, the more likely we would reject H_0 in favor of H_1 (i.e., that the population's regression line has a nonzero intercept). The question that arises is: How "large" does the t statistic have to be to reject H_0? The answer to this question depends on how confident we want to be in our test's result. It is standard practice that researchers settle on 90% or 95% confidence levels or, equivalently, 10% or 5% significance levels (see note 16). Thus, if we choose a significance level of 10% as our cutoff, we can then determine if our t statistic is large enough to reject H_0 for this level. To carry out this test, we can refer to Table C.1 in Appendix C. In the first column of this table we see the heading "df." This is short for **degrees of freedom**, which are calculated by taking our sample size and subtracting from it the number of parameters we have estimated.[18] Thus, for our baseball example, we have a sample size of 32 and we are estimating two parameters (a and b), leaving us with 30 degrees of freedom.[19] Referring to Table C.1, this means we go down to the row labeled 30. Across the top of Table C.1, we see the heading "Confidence Level." Below this, we see "Probability," which is our significance level. Staying with our chosen 10% significance level, this means we go over to the column headed by 0.10 (i.e., 10%). We see, then, for 30 degrees of freedom and a significance level of 0.10, we have the number 1.697 in Table C.1. This number is a **t value** and it represents the minimum value that the absolute value of our t statistic must achieve before we can reject H_0.[20] That is, if

$$|t \text{ statistic}| \geq t \text{ value,}$$

then we can reject H_0, meaning that the intercept is statistically different from 0 at the 10% significance level.

Returning to our baseball example, we have

$$t \text{ statistic} = \frac{2.031}{0.794} = 2.560,$$

which is, in fact, the reported t statistic in Table 3.1 for the intercept, after rounding. Because this t statistic is greater than 1.697, the t value, we can reject H_0 at the 10% significance (90% confidence) level. In fact, we can see that we can achieve even a smaller significance (or higher degree of confidence) with our given t statistic. Our t statistic is larger than the t value of 0.05 (shown as 2.042 in Table C.1) and is larger than the t value of 0.02 (shown as 2.457). However, our t statistic is not larger than the t value of 0.01 (shown as 2.750). Thus, we can reject H_0 at a significance level somewhere between 1% and 2%. What is the exact significance level we can achieve? This is, in fact, what the P value tells us. Referring to Table 3.1, the P value is 0.015759687, implying a significance level of about 1.576% (or, 98.424% confidence level). This illustrates the linkage between the t statistic reported by the program and the associated P value.

We can also use t statistic to perform hypothesis tests other than the zero hypothesis. For example, we saw in our voting regression (see Table 3.2) that the OLS estimation of the slope term, b, yielded a value of approximately 0.871 with a standard error of about 0.184 (rounding to three decimal places). The t statistic and the associated P value show that the population's β is different from 0 at a very small significance level (or, high confidence level). We can, however, ask the following question: Is the population's β different from 1? The value of 0.871 is not too far off from 1 and, given the fact that our value for b is from a sample, it seems possible that if we had the population data and calculated the population regression function we could perhaps end up with β equal to 1. Recall, however, that we cannot simply consider the size of b, but we must consider its size relative to its standard error. Thus, a more formal test is needed. We can state our hypotheses more formally as

$$H_0: \beta = 1,$$
$$H_1: \beta \neq 1.$$

We can perform this test using the t statistic in a similar way as we did before. In this case, however, we need to recalculate the t statistic so as to make it conform to our new hypothesized value of β. We calculate this new t statistic using the following equation:

$$t \text{ statistic} = \frac{b - h}{se(b)},$$ (3.6b)

where b is the OLS sample result for the slope term and h is our hypothesized value for β.[21] Using the values from our voting regression, we have

$$t \text{ statistic} = \frac{0.871 - 1}{0.184} = -0.701$$ (3.6c)

and taking the absolute value for this t statistic, we have 0.701. Turning to Table C.1, we note that there is no row for 48 degrees of freedom (50 observations minus 2 estimated parameters). This is because t tables for all possible values of the degrees of freedom would be too large. Thus, we can use the value 40. Suppose we again choose a level of significance of 10% (or, confidence level of 90%). We see from the table that the t value for 40 degrees of freedom and a significance level of 0.10 is 1.684. Comparing our t statistic to this t value, we see it is less than the t value and so we cannot reject H_0, that β is truly 1 for the population, at the 10% level of significance. In fact, we see that the absolute value of our calculated t statistic is smaller than all but one of the t values, that of 0.681, meaning that we can reject H_0 only at a significance level of about 50% (0.50), which is a *very* weak result. The bottom line is that it is quite possible that β is actually equal to 1.

SUMMING UP

In building regression models, the researcher is responsible for specifying what factors are important in explaining the behavior of a dependent variable. Whether or not the researcher has built a sound model depends on a number of things. Ultimately, though, the model must have some logic to it. In our three examples, we put forth hypotheses about how Y can be explained by the related X variable. Using samples of data, we then calculated the OLS regression for each model and asked the following questions: How well does the model fit the data as a whole? And how do the separate components of the model (i.e., a and b) perform? We have seen that the first question can be answered by using the R^2 value. If the R^2 is "large" (i.e., close to 1), then our model works well; if it is "small" (i.e., close to 0), then our model performs poorly. What value of the R^2 is large enough so that we can claim a good fit is subjective. [22]

With regard to the second question, we have seen that an estimated coefficient's standard error can be used to determine if the coefficient is reliably different from 0. Using the P value (or significance level in the case of SPSS), we have determined that the smaller its value, the more confident we are that the variable included in the model is truly important in explaining the behavior of the dependent variable. We have also seen that we can test other hypothesized values for our parameters by calculating t statistics and comparing them to the t values provided in Table C.1 of Appendix C.

———•◆•———

▼ Notes

1. Again, this will be true only if the CLRM assumptions are met. As we will see in later chapters, our simple models may not satisfy all the necessary conditions that are needed before we can say that we have found the best possible line to fit our data. To continue with our discussion, however, we will assume that all conditions have been satisfied and our least-squares estimates are the best possible.

2. The R^2 is also referred to as the "coefficient of determination."

3. If the R^2 turns out to be equal to 1, then the most likely reason is that the model that was estimated was, in fact, an identity. (See note 4 in Chapter 1.)

4. It should be pointed out that there is no benchmark R^2 value that needs to be achieved before we declare a model to be successful. There are some areas of research where an R^2 of 0.54 would be considered quite good (e.g., models of wage determination in industries other than MLB), whereas in other areas of research it would be considered a weak result (e.g., models that forecast national income). These differences arise for a variety of reasons. It may be due to differences in the availability and quality of data. Or it may simply be the case that some relationships naturally have a larger random component (i.e., u_i) than others.

5. The output shown, as in earlier examples, has been edited to include only the relevant information needed for the present discussion.

6. The column headed "MS" is simply the calculated mean of the sum of squares values. The MS values are not of particular use for us in this book.

7. "Multiple R," reported as approximately 0.5785, is simply the square root of the R^2 and is the absolute value of the correlation between Y and X. This statistic is not often used in our assessment of the model's overall performance.

8. See Gujarati (1995) for details on how the standard errors for the OLS parameter estimates are calculated.

9. Generally speaking, if the absolute value of the estimated parameter (e.g., a or b) is twice (or more) the size of the associated standard error, then the estimated parameter is considered to be a fairly reliable estimate of the population parameter (e.g., α or β).

10. It should be understood that β is a fixed parameter, not a random variable. The value for b from our sample regression, on the other hand, is not a fixed parameter but is a random variable that we use to estimate β.

11. Table 3.2 is similar to Table 3.1, except the prediction errors and predicted values are omitted and other relevant statistics are now included.

12. See Gujarati (1995) for the proof of this result.

13. The probabilities shown under the P-value column are calculated by plugging the given t statistic into the t probability distribution function. This, again, is a complicated matter that goes beyond the scope of this book. Those interested in learning more about the specifics of probability distribution functions are referred to Gujarati (1995) or Greene (2000).

14. If we reject a hypothesis when, in fact, it is true, we are committing what is called a "Type I error." (Accepting a hypothesis when, in fact, it is false is called a "Type II error.") The P value thus represents the probability of committing a Type I error.

15. The "Sig." value of 0.000 represents a rounded figure; it is not perfectly equal to 0, just very close. The fact that the intercept term is significantly different from 0, however, is not of great importance to us because the interpretation of the intercept in this case is not very meaningful.

16. A word on terminology. In determining whether we reject or do not reject the null hypothesis, the P value (or significance level) tells at what probability is the coefficient "significantly" different from 0. For example, a P value of 0.10 means that we can reject the hypothesis that the associated coefficient is 0 at the 10% "significance level." Equivalently, we can say that the null hypothesis is rejected at the 90% "confidence level." The terms significance level and confidence level are used interchangeably and the confidence level is simply 100 minus the significance level.

17. Some older software programs simply provided the standard error for each coefficient estimated and the researcher then had to perform calculations and hypothesis tests by hand. Most software programs now routinely generate t statistics and P values (or significance levels).

18. The meaning of the degrees of freedom is somewhat difficult to explain. The easiest way to think about the degrees of freedom is that it is a measure of how much usable information we have left over from our sample after already performing certain tasks. That is, when we have a sample of data, this data set contains a finite amount of information. If we use part of this information to perform certain calculations, such as estimating a and b in a sample regression function, then we have less information left over to carry out other tasks, such as hypothesis tests. The degrees of freedom, thus, keeps track of how much information we have left for such tasks.

19. Note that both Excel and SPSS report the degrees of freedom. Both programs provide several values for the degrees of freedom, shown as "df," in the ANOVA portion of their output. The relevant values for the purpose of performing t tests are the ones reported for the "Residual." Thus, we see in Table 3.1 a value of 30. In the case of our voting model, with a sample size of 50, in Table 3.3 we have a value of 48.

20. If the value for a or b happens to be negative, then its t statistic will be negative (e.g., the value for b in our abortion regression shown in Table 3.3). The t tables, however, only show positive values. This is because the t distribution is symmetric and centered around 0 and so if we compute a t statistic and find it is negative, we can take the absolute value of this number and compare it to the positive t values shown in Table C.1. For a more detailed discussion of the t distribution, see Gujarati (1995).

21. This equation is very similar to that shown in (3.6a), but with two exceptions. First, we are dealing with b, the slope term, instead of a, the intercept. Second, we are subtracting h in the numerator. On this last point, however, this is only an apparent difference. Recall, in (3.6a), that we were calculating a t statistic for use in testing a hypothesized value of 0 for our population parameter α. In that case, then, h was 0 and so a minus h would simply be equal to a.

22. It should be noted that it is possible for a model to produce a relatively large R^2 while each separate estimated coefficient fails to achieve a sufficient level of significance. It is also possible to achieve a low R^2 while each estimated coefficient is highly significant.

PROBLEMS

3.1 Use Excel (or another program) and the data shown in Table A.6 to calculate the OLS estimation of the following model:

$$Y_i = \alpha + \beta X_i + u_i,$$

where Y_i is hourly wage and X_i is years of education. Interpret the estimated values for α and β. What is the R^2 for this regression? What is its interpretation?

3.2 Use the regression output from Problem 3.1 to perform the following hypothesis tests:

a. H_0: $\alpha = 0$, H_1: $\alpha \neq 0$, at the 5% significance (or, 95% confidence) level.

b. H_0: $\beta = 0$, H_1: $\beta \neq 0$, at the 1% significance (or, 99% confidence) level.

3.3 Use the output shown in Table 3.2 and the t table presented in Appendix C to test the following hypotheses: H_0: $\alpha = 50$, H_1: $\alpha \neq 50$, at the 5% significance (or, 95% confidence) level. [Hint: Use (3.6b) to conduct this test.]

4

MULTIPLE REGRESSION ANALYSIS

The two-variable model we have been working with is the simplest model we can construct. Most things in life, however, are more complicated. Take our baseball example, which hypothesizes that player salaries are a function of how many years a player has played in the major leagues. It should be obvious that this model is too simple. Years of experience may be important in explaining salaries, but clearly there are other factors that are important. In fact, reviewing the data in Appendix A (Table A.1), we see that there are five players with 8 years' experience whose salaries range from $2.75 million (Darryl Hamilton) to $8 million (Sammy Sosa). Years' experience doesn't explain this difference in salary. The data also show that Mike Piazza, who has played 5 years in the major leagues, earned $8 million dollars in 1998, whereas Paul Molitor has played for 19 years but earned only $4.25 million, or about half as much as Piazza. Time on the job may be important, but so are other factors. To improve our model and take into account other important factors, we now begin to discuss the **multiple regression** model. We shall see that, in all three of our examples, a richer model is in order.

INTRODUCTION

Baseball Salaries Revisited

A multiple regression model is simply a model that has two or more independent variables. For example, we can build upon our baseball model to include another measure that theoretically should affect baseball salaries: offensive performance. If we consider players with the same number of years' experience in MLB, then players who perform better offensively may be rewarded with a higher salary, other things being equal. Thus, we can take our original model shown in (1.2b) and add to it, giving us the following model:

$$Y_i = \alpha + \beta_1 X_{1i} + \beta_2 X_{2i} + u_i, \tag{4.1}$$

where Y_i represents player i's salary and X_{1i} is the number of years player i has been in MLB. The variable X_{2i} is the player's career slugging average. Defined as the average number of bases reached per at bat, this variable is a measure of a player's offensive ability. Lastly, u_i is the error term. Notice that we have now added a subscript to the β's and a second subscript to the X's. This is done so that we know, for example, β_1 is the coefficient to the variable X_{1i}. The interpretation of β_1 is essentially the same as it was for our previous model. That is, β_1 shows how player salaries increase, on average, as years of MLB experience increase, other things (such as slugging average) being equal. The value for β_2 shows how player salaries increase, on average, as slugging average increases, other things (such as years of MLB experience) being equal. We would expect a positive value for β_2 because the larger a player's slugging average, the more he contributes to a team's offensive production and as such he should command a higher salary. In sum, each β shows the separate effects of a one-unit increase in its respective X variable on the dependent variable Y, while all other things are held constant. We can take the population regression function shown in (4.1) and write the associated sample regression function as

$$Y_i = a + b_1 X_{1i} + b_2 X_{2i} + e_i, \tag{4.2}$$

where a, b_1, and b_2 are sample estimates for α, β_1, and β_2, respectively.

Given data on Y_i, X_{1i}, and X_{2i}, our goal, then, is quite similar to the two-variable case: We want to find values for a, b_1, and b_2 such that the sum of the squared errors is as small as possible. Formally, we have the following equation for the sum of the squared errors:

$$\sum_{i=1}^{n} e_i^2 = \sum_{i=1}^{n} (Y_i - a - b_1 X_{1i} - b_2 X_{2i})^2. \tag{4.3}$$

We can see that (4.3) is very similar to (2.2), except we now have a second X variable on the right-hand side. At this point, as we saw in Chapter 2, we now essentially face a calculus problem where we would find values of a, b_1, and b_2 that minimize the sum of squared errors. The details of the solution to this calculus problem are somewhat complicated and need not concern us here.[1] Suffice to say that the solution would yield formulas for calculating a, b_1, and b_2 that we could use to define an OLS regression line. Fortunately, software programs such as Excel and SPSS can carry out such calculations for us.

The regression model shown in (4.1) takes our original two-variable model one dimension further and allows us to consider differences in offensive ability. Even this model, however, may not adequately explain differences in salary for players who have the same years of MLB experience and the same offensive ability. Another important factor in determining salary may be defensive ability (i.e., the ability to prevent the opposing team from scoring). For example consider two players with the same years of experience and the same offensive ability (i.e., slugging average). If one player is a better defensive player than the other, he would likely command a higher salary. Thus, we can add to the model shown in (4.1) a variable that captures differences in defensive ability. For example, we may include fielding percentage, which measures a player's ability to make a defensive play without committing an error.[2] Doing so gives us the following sample regression function:

$$Y_i = \alpha + \beta_1 X_{1i} + \beta_2 X_{2i} + \beta_3 X_{3i} + u_i, \tag{4.4}$$

where X_{3i} is player i's fielding percentage and β_3 shows the separate effect of an increase in a player's fielding percentage on his salary, other things being equal. We would expect a positive value for β_3 because better defensive players (i.e., players with larger fielding percentages) should earn higher salaries, other things being equal. The associated sample regression function would be

$$Y_i = a + b_1 X_{1i} + b_2 X_{2i} + b_3 X_{3i} + e_i. \tag{4.5}$$

Our goal now is to find values of a, b_1, b_2, and b_3 that would minimize the sum of squared errors and give us the best-fitting equation for our data. Specifically, we can augment (4.3) to include our third X variable, giving us the following equation for the sum of squared errors:

$$\sum_{i=1}^{n} e_i^2 = \sum_{i=1}^{n} (Y_i - a - b_1 X_{1i} - b_2 X_{2i} - b_3 X_{3i})^2. \tag{4.6}$$

Again, applying differential calculus methods to (4.6), we can find the OLS values for a, b_1, b_2, and b_3. Both Excel and SPSS are capable of performing these complicated calculations quite easily. For example, using Excel to estimate the sample regression model shown in (4.5) yields the results shown in Table 4.1, which provides us with a great deal of information. We can see under the column headed "Coefficients" values given for the intercept, YEARS, SLUGGING, and FIELDING. These values are the OLS estimates for a, b_1, b_2, and b_3, respectively. Thus, the estimated sample regression function (after rounding to three decimal places) is

$$\widehat{Y_i} = -76.234 + 0.291 X_{1i} + 0.022 X_{2i} + 0.695 X_{3i}. \tag{4.7a}$$

Replacing the Y and X's with variable names, we have

$$\begin{aligned}\widehat{\text{SALARY}}_i = {}& -76.234 + 0.291(\text{YEARS}_i) + 0.022(\text{SLUGGING}_i) \\ & + 0.695(\text{FIELDING}_i). \end{aligned} \tag{4.7b}$$

Interpreting (4.7b), we see that the intercept term is a very large, negative number. Technically, this means that a player with zero years' experience, zero slugging average, and zero fielding percentage would earn, on average, $ − 76.23 million! Obviously, this value for a is meaningless as a player realistically cannot earn a negative salary. This result for a exemplifies the point made in Chapter 1 that the intercept does not always have a sensible interpretation.

Moving on to the coefficient for YEARS (i.e., b_1), we see that, on average, a player's salary increases by approximately $0.291 million (or, approximately $291,000) for each additional year of MLB experience, other things being equal. The coefficient for SLUGGING (i.e., b_2) tells us that, on average, if a player's slugging average increases by one point, (i.e., 1/10 of a percent) his salary rises by approximately $0.022 million (or about $22,000), all else considered. Finally, the coefficient for FIELDING (i.e., b_3) tells us that, on average, if a player's fielding percentage increases by 1 percent, his salary is expected to increase by about $0.695 million (or $695,000).

TABLE 4.1

Summary Output

Regression Statistics

Multiple R	0.81968142
R Square	0.67187763
Adjusted R Square	0.636721662
Standard Error	1.674510571
Observations	32

ANOVA

	df	SS	MS	F	Significance F
Regression	3	160.7637621	53.58792068	19.11133911	6.05257E − 07
Residual	28	78.51159831	2.803985654		
Total	31	239.2753604			

	Coefficients	Standard Error	t Stat	P value
Intercept	−76.23398562	24.66415923	−3.090881181	0.00448813
YEARS	0.290659321	0.066903021	4.344487227	0.000165835
SLUGGING	0.022151184	0.004662842	4.750575814	5.48454E − 05
FIELDING	0.694846331	0.249833078	2.781242322	0.009577761

Our results in Table 4.1 were, of course, derived from a sample of data. Thus, we face the same issues raised earlier in Chapter 3 regarding the reliability of the values of a, b_1, b_2, and b_3 as estimates of the population parameters α, β_1, β_2, and β_3. That is, are the values for the estimated parameters statistically different from 0? We can answer this question by referring to the P values provided for each estimated coefficient. As we can see, each estimated coefficient has a P value that is quite small. In fact, the largest one shown for the three b's is only 0.0096 (rounding to four decimal places).[3] What this means is that we can reject the hypothesis, H_0: $\beta = 0$ in favor of H_1: $\beta \neq 0$ at a very high confidence level (or, equivalently, a very small significance level) for all three β's. In other words, there is strong evidence that YEARS, SLUGGING, and FIELDING are significant determinants of a player's salary.[4]

Table 4.1 also provides us with the R^2, reported as approximately 0.672, which we can use to evaluate the overall goodness of fit. In this case, the R^2 tells us that about 67.2% of the variation of salaries can be explained by the variables YEARS, SLUGGING, and FIELDING. Or, more simply, our model explains about two thirds of the behavior of player salaries. Referring to Table 3.1, we can recall that our simple two-variable model yielded an R^2 of approximately 0.335, or 33.5%. Thus, the R^2 has achieved a marked improvement by adding SLUGGING and FIELDING to our model. A word of caution, however: *We cannot directly compare the R^2 from our two-variable model to that of our multiple regression model.* In comparing the R^2's, there are essentially two rules. First, it is only appropriate to directly compare R^2's from two different models if the dependent variable (Y) is the same in both models *and* the number of independent variables (X's) is the same in both models.[5] Second, if the dependent variables are the same, but the number of X variables is not the same, then each R^2 must be adjusted before we can compare them. This adjustment is needed because of a minor problem with the R^2 measure. It can be shown that the R^2 is a nondecreasing function of the number of X variables that appear in our model, regardless of their importance.[6] That is, the R^2 can be artificially inflated by simply adding more X variables to our model, even if adding those X variables is not theoretically justified. The **adjusted R^2** simply takes the regular R^2 and adjusts it downward depending on how many X variables we have. Virtually all software programs that perform regression analysis, including Excel and SPSS, report both the R^2 and the adjusted R^2. Looking at Table 4.1, we see that below the R^2 entry the adjusted R^2 (i.e., "Adjusted R Square"[7]) is reported as approximately 0.637. Referring to Table 3.1, we can see for our two-variable model that the adjusted R^2 for that regression is given as approximately 0.313. Thus, based on adjusted R^2's, the multiple regression model performs better than our simple two-variable model.

In dealing with a multiple regression model, a test is commonly carried out to consider the significance of the model *as a whole*. This kind of test becomes important when we may have somewhat high P values (i.e., small t statistics) for our estimated b's, indicating that our X variables, separately, may not be statistically important.[8] In this event, we may question whether our model, in fact, has anything to say about the dependent variable, Y. It may be the case, however, that, despite weak results for the separate X variables, taken together as a group the X variables are jointly important. Or, in other words, each piece of our model may be weak, but taken as a group they may be strong. Writing this as a formal hypothesis, we have

$$H_0: \beta_1 = \beta_2 = \beta_3 = 0.$$

That is, we have the hypothesis that all the population parameters are truly and jointly equal to 0, meaning that our model as a whole has no ability to explain the behavior of Y.[9] As for the alternative hypothesis, it would simply be that one or more of the β's are not simultaneously equal to 0.[10] This test of the overall significance of a regression model is called the F **test of significance**, the details of which are somewhat complicated. Fortunately, however, most software programs routinely provide the necessary information, making this test quite easy. Referring to Table 4.1, we see under the "ANOVA" heading a column headed "F" (short for F **statistic**) with the value 19.11 (rounding to two decimal places). Next to this column, we see another headed "**Significance F**" with the value 6.05257E − 07 (which uses scientific notation).[11] The Significance F is analogous to the P value, which was used to assess the significance of each X variable separately, except the Significance F is used to assess the significance of the X variables taken as a group. The Significance F tells us the probability that H_0 is true, given our sample results for the regression. In this case, we see that this is an extremely small probability, equal to approximately 0.000000605. In other words, we can be quite confident that the model has something to say about the behavior of Y.

Presidential Elections Revisited

Returning to our presidential election example, recall that our two-variable regression model hypothesized that the percentage of the two-party votes received by the incumbent-party candidate (Y) was a function of the real growth of the economy during the period before the election (X). While real growth was found to be an important predictor for the votes received, there are other economic measures that we can consider. For example, inflation, defined as the rise in the average price level, is an economic evil because it erodes the value of our income, savings, and other assets that are denominated in dollars. Thus, even if real incomes have grown, a high rate of inflation may bring economic harm to voters who have dollar-denominated assets and thus these individuals may, in turn, be less likely to reward candidates from the incumbent party with votes. Thus, we can consider the following multiple regression model that includes both real growth and inflation:

$$Y_t = \alpha + \beta_1 X_{1t} + \beta_2 X_{2t} + u_t. \tag{4.8}$$

The associated sample regression model is

$$Y_t = a + b_1 X_{1t} + b_2 X_{2t} + e_t, \tag{4.9}$$

where X_{1t} is the real percentage growth rate of GDP over the three quarters prior to the election and X_{2t} is the inflation rate over the 15 quarters prior to the election. We expect b_2, the coefficient to X_{2t}, to be negative, implying that, other things being equal, an increase in inflation prior to the election would reduce the percentage of two-party votes received by the incumbent-party candidate.

The procedure for estimating this model is the same as we saw in the baseball example. We employ the OLS method to find the values for a, b_1, and b_2 so that the sum of squared errors is minimized. Using SPSS to perform this regression, we obtain the results shown in Table 4.2. Reviewing Table 4.2, we can see in the "Model Summary" panel that the model explains more than 60% of the variation in Y, according to the R^2. We can also see that this model outperforms the two-variable one we had before as the adjusted R^2 increased from 0.517 to 0.560 (see Table 3.2). Under the panel headed "ANOVA," we see the F statistic, equal to 13.741, and the associated "Sig." (short for Significance F), which can be used to perform a test of overall significance. According

TABLE 4.2

		Model Summary		
Model	R	R Square	Adjusted R Square	Std. Error of the Estimate
1	0.777[a]	0.604	0.560	4.7223231

[a]Predictors: (Constant), INFLATION, GROWTH.

		ANOVA[b]				
Model		Sum of Squares	df	Mean Square	F	Sig.
1	Regression	612.842	2	306.421	13.741	0.000[c]
	Residual	401.406	18	22.300		
	Total	1014.248	20			

[b]Dependent variable: VOTES.

[c]Predictors: (Constant), INFLATION, GROWTH.

		Coefficients[d]			
		Unstandardized Coefficients			
Model		B	Std. Error	t	Sig.
1	(Constant)	54.339	1.964	27.668	0.000
	GROWTH	0.700	0.203	3.450	0.003
	INFLATION	−0.553	0.326	−1.694	0.108

[d]Dependent variable: VOTES.

to the "Sig." value shown, we can reject the hypothesis H_0: $\beta_1 = \beta_2 = 0$ at a very small level of significance (or, equivalently, at a very high level of confidence).[12] That is, we have a statistically significant regression.

The OLS values for a, b_1, and b_2 are given in the third panel headed "Coefficients." Using these values, we can write the equation for the predicted percentage of two-party votes received by the incumbent-party candidate as

$$\widehat{Y}_t = 54.339 + 0.700X_{1t} - 0.553X_{2t}. \qquad (4.10a)$$

Or, using variable names, we have

$$\widehat{\text{VOTES}}_t = 54.339 + 0.700(\text{GROWTH}_t) - 0.553(\text{INFLATION}_t). \quad (4.10b)$$

The coefficient to GROWTH is positive and has a "Sig." value of 0.003, meaning that it is significantly different from 0 at better than the 1% level of significance (or, 99% level of confidence). Its value of 0.700 tells us that a 1% increase in the growth (as measured here) increases the percentage of two-party votes received by the incumbent-party candidate by about 0.7%, all else being equal. The coefficient to INFLATION is negative and has a "Sig." value of 0.108, meaning that the estimated coefficient is statistically different from 0 at only the 10.8% level of significance (or, 89.2% level of confidence). The coefficient's value of −0.553 means that if inflation (as measured here) increases by 1%, the percentage of two-party votes received by the incumbent-party candidate falls by approximately 0.553%, other things considered. This last comment, however, is made with caution due to the somewhat weak results for the "Sig." value for this coefficient. Recall that a "Sig." value of 0.108 means that there is a 10.8% chance of obtaining a $b_2 = -0.553$ when, in fact, the true population's $\beta_2 = 0$. This significance level is larger than the 10% cutoff that is commonly used. Ultimately, it is up to the individual researcher or reader to judge whether this result is too weak or not.[13]

Abortion Rates Revisited

The two-variable model of abortion rates that we estimated in Chapter 3 performed poorly as the R^2 was only 0.016 (see Table 3.3). Recall that the coefficient RELIGION was negative, as expected, but was significantly different from 0 at only the 38.6% level of significance (or, 61.4% confidence level). By any standard, this result is weak, suggesting that there is a strong chance that the religious makeup of a state (as measured by RELIGION) tells us nothing about a state's abortion rate. Clearly, this two-variable model is inadequate.

We can imagine many factors that may affect the abortion rate in a state, beyond its religious breakdown. For example, the field of economics tells us that several factors affect the demand for a particular good or service.[14] In particular, the price of the service should be an important factor. That is, other things being equal, as the price of an abortion increases, the demand for the service should decrease.[15] In addition, income that individuals have to spend on abortions may also affect the demand for abortion services. Specifically, controlling for other factors (i.e., price and religion), as incomes rise, the service becomes more affordable and thus the demand for it increases. Yet, with regard to the effects of income, another competing theory arises. As incomes rise, all else constant, the ability of a parent to be able to afford to care for a child also increases and thus the demand for abortion services may actually *fall*. Which theory for income is correct? We cannot say at the outset and we allow the regression results to give us guidance.

Another factor that may affect the demand for abortion services is antiabortion activity. As we know, abortion is a divisive, socially sensitive issue. It has prompted some individuals to join together in opposing the provision of such services. In fact, many clinics that provide abortion services have been picketed by groups of individuals who oppose abortion. This being the case, in states where this kind of activity occurs frequently, one would expect that the abortion rate may be lower as both providers and consumers of the service face opposition, all else being equal. To control for this factor in our multiple regression model, a variable called PICKETING is included and is equal to the percentage of clinics in a state that have reported experiencing picketing (including contact with patients and/or providers). Other things being equal, we expect that states with a higher value for PICKETING will have a lower abortion rate.

Given the previous discussion, we can write our sample regression function for abortion as follows:

$$Y_i = a + b_1 X_{1i} + b_2 X_{2i} + b_3 X_{3i} + b_4 X_{4i} + e_i, \tag{4.11}$$

where Y_i is the abortion rate in state i, X_{1i} is the value of RELIGION in state i, X_{2i} is the average PRICE of an abortion in state i, X_{3i} is the value of average INCOME in state i, and X_{4i} is the value for PICKETING in state i.[16] Given the previous discussion, we expect negative signs for b_1, b_2, and b_4. As for b_3, we have competing theories and so it could be positive or negative.

TABLE 4.3

Model Summary				
Model	R	R Square	Adjusted R Square	Std. Error of the Estimate
1	0.734[a]	0.539	0.498	7.1272645

[a]Predictors: (Constant), PICKETING, PRICE, RELIGION, INCOME.

ANOVA[b]						
Model		Sum of Squares	df	Mean Square	F	Sig.
1	Regression	2671.720	4	667.930	13.149	0.000[c]
	Residual	2285.905	45	50.798		
	Total	4957.626	49			

[b]Dependent variable: ABORTION RATE.

[c]Predictors: (Constant), PICKETING, PRICE, RELIGION, INCOME.

Coefficients[d]					
		Unstandardized Coefficients			
Model		B	Std. Error	t	Sig.
1	(Constant)	−5.883	9.185	0.641	0.525
	RELIGION	5.058E − 04	0.083	0.006	0.995
	PRICE	−4.51E − 02	0.022	−2.043	0.047
	INCOME	2.389E − 03	0.000	6.196	0.000
	PICKETING	−10.882	3.998	−2.722	0.009

[d]Dependent variable: ABORTION RATE.

Table 4.3 shows the OLS estimation for this multiple regression model. The reported R^2 of 0.539 tells us that our model explains about 53.9% of the variation in abortion rates across states. The adjusted R^2 is shown as 0.498, whereas for the two-variable model it was −0.005.

Thus, we clearly have a better fit in this case.[17] The F statistic (13.149) and the associated "Sig." value (0.000), shown in the ANOVA panel, indicate that the regression as a whole is statistically significant at better than the 0.1% significance (or, 99.9% confidence) level. Using the OLS coefficients shown for our model, we can write the following predicted equation (after rounding):

$$\widehat{Y}_i = -5.883 + 0.0005X_{1i} - 0.0451X_{2i} + 0.0024X_{3i} - 10.882X_{4i}. \quad (4.12a)$$

Rewriting using variable labels, we have

$$\widehat{\text{ABORTION}}_i = -5.883 + 0.0005(\text{RELIGION}_i) - 0.0451(\text{PRICE}_i)$$
$$+ 0.0024(\text{INCOME}_i) - 10.882(\text{PICKETING}_i). \quad (4.12b)$$

Of the four coefficients to the X variables, the one for RELIGION clearly has the wrong sign. It is positive, whereas we hypothesized that it should be negative. This point, however, becomes moot in light of the fact that the small t statistic (0.006) and the large associated "Sig." value (0.995) provide strong evidence that RELIGION is not statistically important in this regression. This outcome illustrates an important point. Namely, in cases where an estimated coefficient is not statistically different from 0 at an acceptable level of significance, interpretation of the estimated coefficient's sign and size is often ill advised because it may imply to the uninformed reader that the variable is statistically important, when, in fact, it is not.

The coefficient to PRICE, on the other hand, has the expected negative sign and it is statistically different from 0 at the 4.7% level of significance. This result supports the hypothesis that, other things considered, as the price of an abortion increases from one state to another, the abortion rate falls. Specifically, a one-dollar increase in PRICE tends to reduce the abortion rate by 0.045, all else being equal.

As for INCOME, recall that we had competing theories as to whether it would have a positive or negative affect on the abortion rate. The estimated coefficient is found to be positive and significantly different from 0 at better than the 0.1% level. This result thus lends support for the theory that, other things being equal, higher incomes make abortions more affordable and thus cause an increase in the abortion rate.[18] That is, according to our results, as incomes rise by one dollar (from one state to another), the abortion rate tends to increase by 0.0024.

Finally, the coefficient for PICKETING has the expected negative sign. The t statistic of -2.772 and "Sig." value of 0.009 show that this estimated coefficient is statistically different from 0 at better than the 1% significance (or, 99% confidence) level. These results support the hypothesis that antiabortion activities (as measured here) have worked to reduce the demand for abortion services, other things considered. The estimated coefficient predicts that a 1% increase in reported PICKETING (from one state to another) leads to a 10.882 drop in the abortion rate, all being else equal.

Further Considerations
for the Multiple Regression Model

For each of our three examples, we have successfully improved our ability to predict the behavior of the dependent variable by building a better, albeit more complicated multiple regression model. There are, however, two additional concerns that arise in the context of the multiple regression model.

The first concern has to do with the limitation of our data set and the ability to add X variables to our model. The limitation is that the number of X variables we include in our model must be less than the number of observations we have in our sample. To illustrate this restriction, consider our baseball example. We expanded the model to include two additional X variables (SLUGGING and FIELDING) and this new multiple regression model outperformed the two-variable model we had before. Given this improvement, we could consider adding other variables that may be important in explaining player salaries. For example, we could consider a player's height and weight or a player's on-base percentage. In fact, there may be many other variables we could consider using in our model. Each time we add another X variable to our model, however, we pay a price in terms of degrees of freedom.[19] For each X variable we add to our model, we must calculate another coefficient, leaving fewer remaining degrees of freedom to perform other tasks such as hypothesis testing. Thus, given that our data set has a limited amount of information, this puts a limit on the number of tasks we can perform.

The second concern has to do with the relationship (if any) among the X variables. Simply put, the OLS method will not tolerate any *perfect, linear* relationship among the X variables in our model. That is, in the case of the multiple regression model, we now add to our list of assumptions outlined in Chapter 2, namely: OLS assumes that there does not exist any perfect linear relationship between the independent variables. That is, we assume there is no perfect **multicollinearity**.[20] To understand this point, consider our baseball example. Suppose we wanted to include a player's age (years and months) as an independent variable with the hypothesis that older players may be rewarded for their experience and leadership they bring to the team. In addition, we can consider when a player was born (year and month) as an independent variable with the hypothesis that players from different generations may have different expectations on how they should be paid for playing baseball. Although these two X variables seem to capture different effects (the "experience factor" and the "generation difference"), they are, in fact, perfectly redundant. That is, if we know when a player was born, then we know his age. These two variables are perfectly linearly related: Age equals current year and month minus birth year and month. In this case, the OLS method cannot distinguish any separate effects of these two variables because they contain the exact same information.[21] The solution: Only one of these two variables can remain in our model; the other must be dropped.

SUMMING UP

The two-variable model, while instructive, is in most cases insufficient and a multiple regression model will be required.[22] We have seen in this chapter that by including relevant variables in our three examples, the estimated multiple regressions outperformed their two-variable counterparts, using the adjusted R^2 as our measure of improvement. The next chapter continues with the development of the multiple regression model by considering not the number, but the various *types* of X variables we can consider.

———•◆•———

▼ Notes

1. The interested reader can find the details for calculating a, b_1, and b_2 in Gujarati (1995).

2. See the footnote to Table A.1 for a definition of fielding percentage.

3. We will ignore the intercept term because it has no real meaning in this regression.

4. Note that, as we saw in Chapter 3, we can work directly with the t statistics to perform zero hypothesis tests as well as testing nonzero hypotheses.

5. For example, we could compare the R^2 from the two-variable model estimation shown in Table 3.1 to the R^2 found from a sample regression where salaries are modeled as a function of slugging average *instead* of years in MLB. In this case, the dependent variable would be the same (player salaries) and both models would have just one X variable. In fact, running this regression (salaries as a function of slugging average alone) yields an R^2 of approximately 0.293. This R^2 is less than the one found in Table 3.1 where it was reported as approximately 0.335. Thus, in comparing these two models, we prefer the one that uses years in MLB. As we shall see, however, both of these two-variable models are inferior to the multiple regression model shown in (4.4).

6. For proof of this fact, see Gujarati (1995), page 207.

7. The adjusted R^2 is sometimes denoted by putting a "bar" above the regular R^2, i.e., $\overline{R}^2$.

8. This is clearly not the case for our baseball example as all the P values are quite small, indicating that all three X variables are statistically important determinants of Y.

9. This kind of test is unnecessary for the two-variable model we saw in earlier chapters. This is because in the two-variable case there is only one β and so the simple P-value test is sufficient.

10. Although the R^2 tells us in some sense how well our model is performing as a whole, it is not a statistical test per se. It can be shown, however, that testing the hypothesis, $H_0: \beta_1 = \beta_2 = \beta_3 = 0$, is equivalent to testing the hypothesis, $H_0: R^2 = 0$ [see, e.g., Gujarati (1995), page 248].

11. The F statistic follows an F distribution which, in turn, can be used to generate the Significance F. For more details on F distributions and the F statistic, see Gujarati (1995).

12. The "Sig." value shown as 0.000 is rounded to three digits and means that we can reject $H_0: \beta_1 = \beta_2 = 0$ at the less than one tenth of 1% level of significance.

13. Some researchers, in fact, insist on 5% as the cutoff before declaring a coefficient statistically different from 0.

14. I do not wish to trivialize the socially important issue of abortion by considering it as a service that is bought by a consumer. Thinking in this way simply allows us to consider economic factors that may have an impact on abortion rates.

15. Economists call this inverse relationship between price and the quantity demanded the "law of demand."

16. See the footnote to Table A.4 for a description of these variables.

17. A negative adjusted R^2 can occur in cases where we have a particularly poor fit as was the case in the two-variable model for abortion rates.

18. There is, in fact, another interpretation for the positive effect of income on the abortion rate. Economists use the term "opportunity costs" when discussing what one gives up in order to pursue something else. In this case, to carry a pregnancy to term and raise a child, it is likely that the parents would have to sacrifice part of their income. This being the case, individuals with higher incomes would face a higher opportunity cost of being parents than those with lower incomes and thus the higher-income individuals may be more inclined to terminate a pregnancy.

19. See note 18 in Chapter 3.

20. Lesser degrees of multicollinearity, which is the case where two or more X variables are less than perfectly linearly related, is not necessarily a problem. We will have more to say about this case in Chapter 6. Furthermore, perfect *nonlinear* relationships are not ruled out, as we will discuss in Chapter 5.

21. In fact, any attempt to estimate a model that has a perfectly multicollinear relationship will generally cause the software program to return an error message.

22. In fact, if we estimate only a two-variable model and in so doing exclude other relevant variable(s), then we have committed a model specification error. In this case, we cannot be sure that the OLS-estimated model is BLUE (see the discussion on the CLRM assumptions in Chapter 2). For more on specification errors, see Gujarati (1995), page 455.

PROBLEMS

4.1 Using the data in Table A.4 of Appendix A, reestimate the model shown in (4.11) but now include the variable EDUC as an independent variable. Using your output, answer the following questions:

    a. What is the interpretation of the estimated coefficient to EDUC?

    b. Does this model, as a whole, outperform the one without EDUC? Explain.

    c. Is the coefficient to EDUC statistically different from 0 at the 5% level of significance (95% level of confidence)?

4.2 Using the output from Table 4.2, test whether the coefficient to GROWTH is statistically different from 1.0 at the 10% significance (90% confidence) level. [Hint: Use (3.6b) to answer this question.]

4.3 Using (4.7b), predict the salary of a MLB player who has 10 years of experience, a slugging average of 380, and a fielding percentage of 97.5.

4.4 Returning to our model of wages as described in Problem 3.1, suppose we now add experience to the equation. That is, we have

$$Y_i = \alpha + \beta_1 X_{1i} + \beta_2 X_{2i} + u_i,$$

where Y_i is hourly wage and X_{1i} is years of education and X_{2i} is years of work experience.

a. Use SPSS or Excel (or another program) to perform an OLS estimate of this multiple regression model and interpret the estimated coefficients.

b. Test whether the estimated values for β_1 and β_2 are statistically different from 0 at the 5% level of significance.

4.5 Recall our model of SUV prices in Problem 2.3. Suppose we now add engine size as an independent variable to our model, giving us

$$Y_i = \alpha + \beta_1 X_{1i} + \beta_2 X_{2i} + u_i,$$

where Y_i is MSRP, X_{1i} is horsepower, and X_{2i} is engine size (in liters).

a. Use SPSS or Excel to calculate the OLS estimate of this model and interpret the results. Are both estimated slope coefficients statistically different from 0 at the 5% level of significance? Explain.

b. Given that the R^2 from the model without engine size is 0.491 with an adjusted R^2 of 0.477, does this model with engine size produce a better fit? Explain.

5

NONLINEAR, DUMMY, INTERACTION, AND TIME VARIABLES

In the previous chapter, we built multiple regression models for our three examples. In each case, new independent (X) variables were introduced that we hypothesized were important in explaining the dependent variable (Y). In all three examples, the new X variables were continuous measures and they entered our model linearly (i.e., with the power of 1). There are, however, other *types* of variables we can consider. Four types that we will discuss in this chapter are nonlinear X variables; **dummy variables**, which denote "categories;" **interaction variables**, which show how two variables may interact with each other, and a variable called a **time index**, which allows us to study time trends.

INTRODUCTION

Nonlinear Independent Variables

In Chapter 1, where we introduced our baseball example, we hypothesized that a player's salary may increase with the number of years he has played MLB [see (1.2b)]. The theory is simply that a player's skills improve the longer he plays in the major leagues. As a result, he is rewarded with increased pay. While this may be true, it may also be the case that the reward to time on the job is not constant throughout a player's career. That is, a player may witness larger salary increases in the early years, but smaller increases in the later years. This may be the case because in the beginning of a player's career experience is

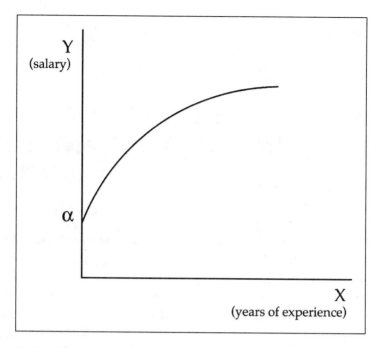

Figure 5.1.

very important as he "learns the ropes" and improves rapidly. Later in a player's career, additional experience may improve his performance, but to a lesser degree because the player is already quite experienced.[1] Mathematically, this would imply a nonlinear function such as

$$Y = \alpha + \beta_1 X + \beta_2 X^2. \qquad (5.1)$$

The function shown in (5.1) is called a quadratic and, if β_1 is positive and β_2 is negative, Y will increase as X increases, but at a decreasing rate. This kind of function would produce a graph like the one shown in Figure 5.1.[2] As shown in Figure 5.1, salary increases with years of experience, but at a decreasing rate. Thus, all that is needed to allow for such a nonlinear relationship in our regression model is to create a new variable that is equal to the square of YEARS.[3] This new variable can now be included with the other X variables we had in our sample regression function as shown in (4.5). This gives us a new sample regression function:

$$Y_i = a + b_1X_{1i} + b_2X_{2i} + b_3X_{3i} + b_4X_{1i}^2 + e_i, \qquad (5.2)$$

where the X variables are the same as before: X_{1i} is years of MLB experience, X_{2i} is a player's career slugging average, and X_{3i} is a player's fielding percentage. Notice that we have now added X_{1i}^2 to the list of independent variables. Before going any further, it should be pointed out that including the square of an existing X variable as an additional independent variable does not violate the assumption made in Chapter 4 of no perfect multicollinearity. As noted before, this assumption rules out perfect *linear* relationships, but not perfect *nonlinear* relationships.[4]

Using Excel to calculate the OLS estimates for a, b_1, b_2, b_3, and b_4, we obtain the results shown in Table 5.1. We see in Table 5.1 that the R^2 is about 0.831. This tells us that our model explains about 83.1% of the variation in salaries. Comparing the results shown in Table 5.1 to those in Table 4.1, we see that the adjusted R^2 has substantially increased with the inclusion of years of MLB experience squared (shown as YEARS SQUARED in Table 5.1). Furthermore, the P value for the coefficient to YEARS SQUARED is approximately $2.793E - 05$ (using scientific notation), which tells us that the b_4 is statistically different from 0 at a very small significance level (i.e., high confidence level). In sum, inclusion of YEARS SQUARED has improved our model. Using the value of the estimated coefficients from Table 5.1, the predicted equation, after rounding, is

$$\widehat{Y}_i = -27.681 + 1.177X_{1i} + 0.017X_{2i} + 0.195X_{3i} - 0.048X_{1i}^2. \qquad (5.3a)$$

Or,

$$\widehat{SALARY}_i = -27.681 + 1.177(YEARS_i) + 0.017(SLUGGING_i)$$
$$+ 0.195(FIELDING_i) - 0.048(YEARS_i^2). \qquad (5.3b)$$

We see that (ignoring the constant term) all of the coefficients in (5.3b) have the expected signs. SLUGGING is positive and significantly different from 0, a very small significance level (i.e., high confidence level). Whereas FIELDING was statistically significant in the model shown in (4.5), it is no longer statistically significant at an acceptable level here.

TABLE 5.1

Summary Output

Regression Statistics

Multiple R	0.911417957
R Square	0.830682692
Adjusted R Square	0.805599647
Standard Error	1.224948372
Observations	32

ANOVA

	df	SS	MS	F	Significance F
Regression	4	198.7619005	49.69047513	33.11597758	$4.72426E-10$
Residual	27	40.51345986	1.500498513		
Total	31	239.2753604			

	Coefficients	Standard Error	t Statistic	P Value
Intercept	−27.68060297	20.46027903	−1.352894695	0.187310386
YEARS	1.176862877	0.18277855	6.438736254	$6.7158E-07$
SLUGGING	0.017007862	0.003560827	4.776380394	$5.55534E-05$
FIELDING	0.19484304	0.208022424	0.936644406	0.357244516
YEARS SQUARED	−0.048182928	0.009574802	−5.032263524	$2.79266E-05$

The coefficients to YEARS and YEARS SQUARED are both statistically different from 0 at a very low level of significance (i.e., high level of confidence). The combined effect of YEARS and YEARS SQUARED supports the hypothesis described earlier that SALARY tends to increase, but at a declining rate, as a player's years in MLB increase. As an example, consider a player with 2 years of MLB experience. We can use the results shown in (5.3a) to calculate the estimated increase in his salary for an additional year of experience. To do so, we first plug into (5.3a) the value of 2 for X_{1i}, and 4 for X_{1i}^2, then we multiply these values by the estimated coefficients for b_1 and b_4 and then add these results together. This is then repeated for X_{1i} equal to 3 and X_{1i}^2 equal to 9. The difference between these two values is the gain in salary due to a third year of experience. Performing these calculations, we find that the expected increase in pay to this player for an additional year of experience would be approximately $0.937 millions (or $937,000), other things constant. On the other hand, using a similar calculation, we find that a player with 5 years of MLB experience can expect a salary increase of about $0.649 millions (or $649,000) for an additional year of experience, other things constant.[5] As expected, gains decrease with greater experience.

Including the square of YEARS in our model has clearly improved our model's overall performance. It should be noted that we are not restricted to only adding square terms. In some cases, we may wish to add the square of a variable *and* its cube when we hypothesize more complicated nonlinear relationships.[6] In addition, some nonlinear relationships can be "straightened out" by transforming existing variables.[7]

Dummy Independent Variables

In some cases, one or more of our independent variables may not be a continuous measure. For example, we can consider our presidential election model. One factor that we have not considered is whether the incumbent-party candidate is, in fact, the incumbent himself. An example would be Bill Clinton, who was the incumbent-party candidate in 1996 when he ran for reelection.[8] We can contemplate how being the incumbent affects the percentage of two-party votes received by the incumbent-party candidate. It would seem likely that, other things being equal (including growth of the economy and inflation), the incumbent would have an advantage over another candidate simply because he is better known to the voters. That is, other things being equal, incumbent candidates would, on average, receive a larger percentage of the two-party votes. We can test this hypothesis by *creating* a "dummy variable" that indicates whether the incumbent-party candidate is indeed the president running for reelection.[9] To facilitate a graphical representation of dummy variable effects, we can consider our earlier two-variable sample regression model shown in (1.6), but now including a dummy variable to indicate whether the incumbent-party candidate was, in fact, the incumbent:

$$Y_t = a + b_1 X_{1t} + b_3 X_{3t} + e_t, \tag{5.4a}$$

where X_{3t} takes the value of 1 if the incumbent-party candidate is the president running for reelection, 0 otherwise (we have excluded $b_2 X_{2t}$ for the moment so that our graph will be two dimensional; it will be added back later). Thus, the coefficient b_3 would represent the effect on the Y_t, if any, of being the incumbent. Note that in the case where the incumbent-party candidate is *not* the incumbent, then X_{3t} would be 0 and the predicted portion of (5.4a) would be

$$\widehat{Y}_t = a + b_1 X_{1t}. \tag{5.4b}$$

However, if the incumbent-party candidate is, in fact, the incumbent, then X_{3t} would be equal to 1 and the predicted portion of (5.4a) becomes

$$\widehat{Y}_t = a + b_1 X_{1t} + b_3. \tag{5.4c}$$

As we can see, the only difference between (5.4b) and (5.4c) is that the latter includes b_3. Or, in other words, the intercept value for (5.4b) is a, whereas in (5.4c) it is $a + b_3$. Graphically, these two equations are simply parallel lines with different intercept values. Figure 5.2 shows an example of this kind. Notice that the line for incumbent-party candidates who are, in fact, incumbents is shifted up by the value b_3. This shift represents the advantage of being an incumbent over nonincumbents, for any given level of economic growth (i.e., for all values of X_{1t}).

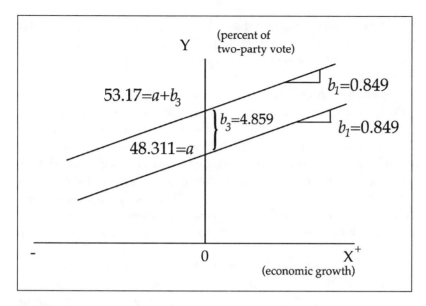

Figure 5.2.

Using OLS to estimate the sample regression function shown in (5.4a) yields the following estimated relationship:

$$\widehat{\text{VOTES}}_t = 48.311 + 0.849(\text{GROWTH}_t) + 4.859(\text{INCUMBENT}_t). \quad (5.4d)$$

According to the preceding estimated equation, given the same level of growth in the economy, incumbents enjoy about a 4.8% advantage over nonincumbents. That is, in terms of Figure 5.2, the sample regression line for incumbents would be shifted vertically upward by about 4.8% above the sample regression line for nonincumbents. This result, however, should not be taken too seriously because for reasons of simplicity we have excluded the independent variable INFLATION, which, as we have previously seen [see (4.10b)], is an important determinant of VOTES.[10]

Adding back to this model the variable for inflation, X_{2t}, we now have

$$Y_t = a + b_1 X_{1t} + b_2 X_{2t} + b_3 X_{3t} + e_t. \quad (5.5)$$

In this case, we would have a three-dimensional graph (which would be difficult to draw and hence not attempted). In this case, however, the interpretation of the coefficient b_3 would be similar as it was in the simpler case discussed previously. Namely, given the same level of economic growth *and* inflation, b_3 represents the added votes that an incumbent candidate is expected to receive as compared to nonincumbents.

It should be noted that X_{3t} is not a continuous measure, such as growth, which can theoretically take on any value. This variable takes on only the values of 1 or 0.[11] Reviewing Table A.3, we see the column headed "Incumbent." The values for this column indicate if the incumbent ran for reelection. A value of 1 indicates yes; 0 indicates no. The treatment for this variable, however, is no different form that for any other X variable in our model. Using SPSS, we simply include this dummy variable on the list of independent variables and perform the OLS estimation as usual. Estimating (5.5), we have the results below in Table 5.2.

Starting with the model summary, we see that R^2 is 0.707, meaning that about 70.7% of the variation in Y is explained by our model. As for the adjusted R^2, it is now 0.665, whereas in the model without the dummy variable INCUMBENT it was 0.560 (see Table 4.2). Thus, based on the adjusted R^2, this model performs better. The F statistic and the associated "Sig." value (shown in the ANOVA panel) indicate that the regression as a whole is statistically significant at a very high level of confidence (i.e., small level of significance).[12]

The estimated coefficients for all of the X variables have the predicted signs. Using these values, we have the following estimated sample regression function:

$$\widehat{Y}_t = 51.077 + 0.686X_{1t} - 0.526X_{2t} + 4.728X_{3t}. \qquad (5.6a)$$

Rewriting using variable names, we have

$$\widehat{\text{VOTES}}_t = 51.077 + 0.686(\text{GROWTH}_t) - 0.526(\text{INFLATION}_t)$$
$$+ 4.728(\text{INCUMBENT}_t). \qquad (5.6b)$$

Checking the significance of each estimated coefficient, we see that the "Sig." values for the constant and GROWTH indicate that they are significantly different from 0 at better than the 1% significance (or, 99% confidence) level. INFLATION, with a "Sig." value of 0.086, is statistically different from 0 at the 8.6% significance (or, 91.4% confidence) level. Concerning our dummy variable, INCUMBENT, it has

TABLE 5.2

		Model Summary		
Model	R	R Square	Adjusted R Square	Std. Error of the Estimate
1	0.841[a]	0.707	0.655	4.1839

[a] Predictors: (Constant), INCUMBENT, GROWTH, INFLATION.

			ANOVA[b]			
Model		Sum of Squares	df	Mean Square	F	Sig.
1	Regression	716.662	3	238.887	13.647	0.000[c]
	Residual	297.586	17	17.505		
	Total	1014.248	20			

[b] Dependent variable: VOTES.

[c] Predictors: (Constant), INCUMBENT, GROWTH, INFLATION.

		Coefficients[d]			
		Unstandardized Coefficients			
Model		B	Std. Error	t	Sig.
1	(Constant)	51.077	2.196	23.262	0.000
	GROWTH	0.686	0.180	3.817	0.001
	INFLATION	−0.526	0.289	−1.820	0.086
	INCUMBENT	4.728	1.941	2.435	0.026

[d] Dependent variable: VOTES.

a "Sig." value of 0.026, indicating that it is statistically different from 0 at the 2.6% significance (or, 97.4% confidence) level. Thus, in sum, the results give strong support for the hypothesis that GROWTH and INCUMBENT are important in explaining VOTES, and reasonably good support that INFLATION is an important determinant for VOTES as well.

The interpretation of the estimated coefficients for the constant, GROWTH, and INFLATION is similar to what it was before. That is, all else being equal, if GROWTH, INFLATION, and INCUMBENT are all 0, then the constant term implies that, on average, the incumbent-party candidate is expected to win about 51.077% of the two-party votes. The coefficient for GROWTH suggests that, on average, a 1% increase in the growth of the economy prior to the election will increase the percentage of two-party votes that the incumbent-party candidate wins by approximately 0.686%, all else being equal. If INFLATION rises by 1%, all else constant, the regression results imply that the incumbent-party candidate will lose about 0.526% of the two-party votes. Finally, the coefficient to our dummy variable INCUMBENT is shown as 4.728. This suggests that, controlling for growth and inflation, if the incumbent-party candidate is, in fact, the incumbent (i.e., $X_{3t} = 1$), then he is expected to receive about 4.728% more of the two-party votes than someone who is the incumbent-party candidate but who is not the incumbent (i.e., $X_{3t} = 0$). In other words, all else being equal, if the president runs for reelection, he has a considerable advantage over someone from the same party running for his or her first term as president.[13]

Dummy variables are widely used in regression analysis.[14] For example, in research projects where a data set has observations on both males and females, one may wish to create a dummy variable to control for gender differences (e.g., a dummy variable that takes the value of 1 if the individual is a male, 0 if a female). Or we can consider the effects of education by creating a dummy variable that is 1 for a high school graduate, 0 otherwise.

In the preceding example, we created a variable that indicated whether something was true where there were only two possibilities: Either the incumbent-party candidate was the incumbent or he was not the incumbent. In some cases, there may be more than two possible categories. For example, in our analysis of salary paid to baseball players, we could consider whether a player's salary depends on his race. In this case, we can consider three categories for a player's race: White, Black, or Non-Black Hispanic. To cover these three categories, we need, however, to create only two dummy variables. One dummy variable takes the value of 1 if the player is Black, 0 otherwise.[15] The second dummy variable takes the value of 1 if the player is a Non-Black Hispanic, 0 otherwise. If both of these dummy variables equal 0, then it must be the case that the player is White. This example demonstrates an important rule regarding dummy variables: *If we have m categories for which we wish to control, we need only m − 1 dummy variables.* This is true because if the values of all $m - 1$ dummy variables are 0, then it must be the case that the observation belongs to the excluded category. In fact, if we tried to include m dummy variables in our model, we would end up violating the regression model assumption of *no perfect multicollinearity* (see Chapter 4) because our model would contain redundant information and software programs would return an error message.

In addition to race, we can also consider whether salaries differ across leagues. That is, is salary higher or lower in the National League as compared to the American League? Given that we have two categories, we can control for league effects by creating one dummy variable, which takes on the value of 1 if a player is in the National League, 0 otherwise (i.e., if a player is in the American League).

Incorporating the two dummy variables for Black and Non-Black Hispanic players and the dummy variable for the National League into our model, the sample regression function becomes

$$Y_i = a + b_1 X_{1i} + b_2 X_{2i} + b_3 X_{3i} + b_4 X_{1i}^2 + b_5 X_{5i} + b_6 X_{6i} + b_7 X_{7i} + e_i, \quad (5.7a)$$

where X_{5i} takes the value of 1 if the player is Black, 0 otherwise. The variable X_{6i} takes the value of 1 if the player is Non-Black Hispanic, 0 otherwise. And X_{7i} is equal to 1 if the player is in the National League, 0 otherwise. In this example, our "base category" for race would be players who are White. For White players, X_{5i} and X_{6i} are both 0 and thus the terms $b_5 X_{5i}$ and $b_6 X_{6i}$ in (5.7a) drop out, leaving the following sample regression function:

$$Y_i = a + b_1 X_{1i} + b_2 X_{2i} + b_3 X_{3i} + b_4 X_{1i}^2 + b_7 X_{7i} + e_i. \quad (5.7b)$$

If a player is Black, then X_{5i} is 1, giving us the following sample regression function:

$$Y_i = a + b_1 X_{1i} + b_2 X_{2i} + b_3 X_{3i} + b_4 X_{1i}^2 + b_5(1) + b_7 X_{7i} + e_i. \quad (5.7c)$$

Notice that the only difference between (5.7b) and (5.7c) is that the latter includes b_5. Thus, the expected effect of being a Black player is to increase (or decrease) a player's salary above (or below) a White player's salary by the amount b_5, other things being equal. If a player is Hispanic, then X_{6i} will be 1, giving us the following sample regression function:

$$Y_i = a + b_1X_{1i} + b_2X_{2i} + b_3X_{3i} + b_4X_{1i}^2 + b_6(1) + b_7X_{7i} + e_i. \qquad (5.7d)$$

In this case, the value of b_6 would represent the difference in salary (positive or negative) for Hispanic players as compared to our base case of White players, all else being equal. Notice, as in the case of our voting model, the dummy variables simply alter the intercept of the model shown in (5.6a). That is, in the case of a White player, the intercept is just a, as b_5X_{5i} and b_6X_{6i} drop out of the equation. In the case of a Black player, b_5 remains in the sample regression function. This is simply a number (whatever b_5 turns out to be) that adjusts the intercept a upward or downward, depending on the sign of b_5. Similarly for the case of Hispanics, b_6 remains in the sample regression function and adjusts the intercept a upward or downward, depending on the sign of b_6. As for the dummy variable for the National League, the analysis is similar. If a player is in the National League, then the term $X_{7i} = 1$ and the term b_7 is added to the intercept and will adjust it upward or downward, depending on the value of b_7. The value of b_7 thus reflects the pay differential (if any) for National League players as compared to the base case, which is the American League, other things being equal.

Using the data in Table A.1 and SPSS to estimate the model shown in (5.7a), we obtain the results shown in Table 5.3. Notice that Table 5.3 reports an R^2 of 0.849, meaning that our model explains about 84.9% of the variation in player salaries. Comparing the adjusted R^2 here to that of Table 5.1, we see there is essentially no increase, which raises the question as to whether these added independent variables are important factors in determining salary (we shall have more to say about this later). The F statistic and associated "Sig." value indicate that, overall, our model is statistically significant. Using the estimated coefficients shown in Table 5.3, we can write our predicted equation as

$$\widehat{SALARY}_i = -27.143 + 1.119(YEARS_i) + 0.017(SLUGGING_i)$$
$$+ 0.187(FIELDING_i) - 0.045(YEARS_i^2) + 0.235(BLACK_i)$$
$$+ 1.495(HISPANIC_i) + 0.249(LEAGUE_i). \qquad (5.7e)$$

TABLE 5.3

		Model Summary		
Model	R	R Square	Adjusted R Square	Std. Error of the Estimate
1	0.921[a]	0.849	0.805	1.2272

[a]Predictors: (Constant), LEAGUE, HISPANIC, YEARS SQUARED, SLUGGING, FIELDING, BLACK, YEARS.

		ANOVA[b]				
Model		Sum of Squares	df	Mean Square	F	Sig.
1	Regression	203.132	7	29.019	19.269	0.000[c]
	Residual	36.143	24	1.506		
	Total	239.275	31			

[b]Dependent variable: SALARY.

[c]Predictors: (Constant), LEAGUE, HISPANIC, YEARS SQUARED, SLUGGING, FIELDING, BLACK, YEARS.

		Coefficients[d]			
		Unstandardized Coefficients			
Model		B	Std. Error	t	Sig.
1	(Constant)	−27.143	21.482	−1.264	0.219
	YEARS	1.119	0.189	5.929	0.000
	YEARS SQUARED	$-4.53E-02$	0.010	−4.584	0.000
	SLUGGING	$1.732E-02$	0.004	4.727	0.000
	FIELDING	0.187	0.220	0.851	0.403
	BLACK	0.235	0.497	0.473	0.641
	HISPANIC	1.495	0.937	1.597	0.123
	LEAGUE	0.249	0.445	0.560	0.581

[d]Dependent variable: SALARY.

Focusing on the coefficients for b_5 and b_6, the first is the coefficient to BLACK, which is positive, indicating that, other things being equal, Black players tend to earn more than Whites. Specifically, a Black player is expected to earn about $0.235 million more than a White player with equal ability (i.e., with the same number of YEAR, YEAR SQUARED, SLUGGING, FIELDING, and LEAGUE values). As for Non-Black Hispanics, they tend to earn about $1.5 million more than White players with equal values for the other independent variables. Finally, the dummy variable for LEAGUE suggests that players in the National League earn, on average, about $0.25 million more than American Leaguers, all else being equal. *While these results are interesting and instructive, the coefficients to BLACK, HISPANIC, and LEAGUE are not statistically different from 0 at a credible significance level, judging by the "Sig." values shown.* Thus, there is no credible evidence that either race or league is a factor in the salary of MLB players. This result is reflected in the adjusted R^2, which, as noted before, is the same as the model that did not include dummy variables for race and league. Remember, however, that this conclusion is based on the sample used in this regression. A different (perhaps larger) sample might produce different results. In any case, as pointed out in Chapter 4, in practice when estimated coefficients are not statistically different from 0 at a credible significance level (e.g., 10% or smaller), it is best not to provide any interpretation of the estimated coefficients because they are statistically meaningless. The interpretation of the estimated coefficients for BLACK, HISPANIC, and LEAGUE are given here simply to illustrate the meaning of the dummy variable coefficients.

Interaction Variables

Our previous voting model, which included a dummy variable for whether or not the incumbent-party candidate was actually the incumbent, implicitly assumed that the only difference between incumbents and nonincumbents is the intercept term [see (5.4a)]. Another possibility, however, is that there may be differences in the slope coefficients between incumbents and nonincumbents. For example, it may be the case that, as the economy grows, an incumbent may be rewarded with votes at a greater *rate* than nonincumbents. The reasoning may be that voters consider the actual incumbent as more responsible for the growth in the country than simply a candidate who is not the current president, but only belongs to the same party as the president.

To capture this effect, consider the following sample regression function (again, we will temporarily exclude the variable for inflation in order to facilitate a graphical representation):

$$Y_t = a + b_1 X_{1t} + b_4(X_{1t} * X_{3t}) + e_t, \qquad (5.8a)$$

where Y_t, X_{1t}, and X_{3t} are as they were in (5.4a), namely, the percentage of two-party votes won by the incumbent-party candidate, the real growth rate, and a dummy variable for whether or not the incumbent-party candidate is actually the incumbent, respectively. In this case, the dummy variable for incumbency does not enter into the equation separately as in (5.4a), but rather it is being multiplied by, or "interacted" with, the variable for growth. This interaction effect captures the difference in the *rate* of votes gained due to economic growth between incumbent and nonincumbent candidates. To see this, consider how (5.8a) changes for incumbents versus nonincumbents. That is, suppose the incumbent-party candidate is not the incumbent (e.g., Al Gore, who is the incumbent-party candidate for the 2000 election, but he is not the incumbent president). Then the predicted relationship for (5.8a) becomes

$$\widehat{Y}_t = a + b_1 X_{1t} + b_4(X_{1t} * 0) = a + b_1 X_{1t}. \qquad (5.8b)$$

We can compare this to the case where the incumbent-party candidate is, in fact, the incumbent (e.g., Bill Clinton running for reelection in 1996):

$$\widehat{Y}_t = a + b_1 X_{1t} + b_4(X_{1t} * 1) = a + b_1 X_{1t} + b_4 X_{1t} = a + (b_1 + b_4)X_{1t}. \quad (5.8c)$$

Comparing (5.8b) and (5.8c), we see that both have the same intercept, a, but the coefficients to variable X_{1t} differ. For nonincumbents, the coefficient to GROWTH is b_1, whereas for incumbents it is $(b_1 + b_4)$. These cases are shown graphically in Figure 5.3.

Before estimating (5.4a), we can first bring back the variable INFLATION (X_{2t}) into our model (and in doing so reduce the risk of misspecifying the model), giving us

$$Y_t = a + b_1 X_{1t} + b_2 X_{2t} + b_4(X_{1t} * X_{3t}) + e_t. \qquad (5.9a)$$

Using SPSS to estimate this equation produces the results given in Table 5.4. Using the results shown in Table 5.4, we can write the estimated

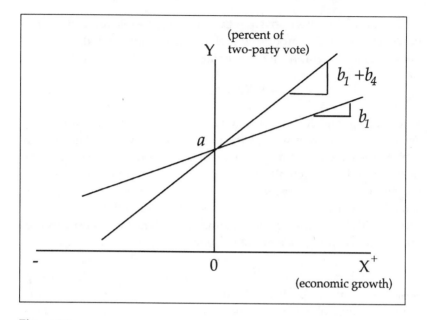

Figure 5.3.

regression as

$$\widehat{\text{VOTES}}_t = 55.148 + 0.397(\text{GROWTH}_t) - 0.717(\text{INFLATION}_t)$$
$$+ 0.34(\text{GROWTH} * \text{INCUMBENT}_t). \qquad (5.9b)$$

Interpreting this result, the intercept predicts that the percentage of the two-party vote received by the incumbent-party candidate is, on average, about 55%, when INFLATION and GROWTH are 0. The coefficient to INFLATION, −0.717, suggests that, other things being equal (i.e., GROWTH), the incumbent-party candidate will receive about 0.7% less of the two-party vote when inflation increases by 1%. Finally, the effect of a 1% increase in GROWTH, holding other things constant (i.e., INFLATION), will increase the percentage of the two-party vote by 0.397% for nonincumbents and by $(0.397 + 0.34) = 0.737\%$ for incumbents. This interpretation is instructive, however, *these results are not reliable*, given, as shown in Table 5.4, that the coefficients to GROWTH and GROWTH∗INCUMBENT are *not statistically different from 0 at a credible level of significance.*[16]

It should be noted that we need not consider the dummy variable (additive) form of our model [i.e., (5.5)] separately from our interaction

TABLE 5.4

<table>
<tr><td colspan="5" align="center">Model Summary</td></tr>
<tr><td>Model</td><td>R</td><td>R Square</td><td>Adjusted R
Square</td><td>Std. Error
of the Estimate</td></tr>
<tr><td>1</td><td>0.784[a]</td><td>0.615</td><td>0.547</td><td>4.7930</td></tr>
</table>

[a]Predictors: (Constant), GROWTH*INCUMBENT, INFLATION, GROWTH.

<table>
<tr><td colspan="7" align="center">ANOVA[b]</td></tr>
<tr><td>Model</td><td></td><td>Sum of
Squares</td><td>df</td><td>Mean
Square</td><td>F</td><td>Sig.</td></tr>
<tr><td>1</td><td>Regression</td><td>623.713</td><td>3</td><td>207.904</td><td>9.050</td><td>0.001[c]</td></tr>
<tr><td></td><td>Residual</td><td>390.535</td><td>17</td><td>22.973</td><td></td><td></td></tr>
<tr><td></td><td>Total</td><td>1014.248</td><td>20</td><td></td><td></td><td></td></tr>
</table>

[b]Dependent variable: VOTES.

[c]Predictors: (Constant), GROWTH*INCUMBENT, INFLATION, GROWTH.

<table>
<tr><td colspan="6" align="center">Coefficients[d]</td></tr>
<tr><td></td><td></td><td colspan="2" align="center">Unstandardized
Coefficients</td><td></td><td></td></tr>
<tr><td>Model</td><td></td><td>B</td><td>Std. Error</td><td>t</td><td>Sig.</td></tr>
<tr><td>1</td><td>(Constant)</td><td>55.148</td><td>2.315</td><td>23.827</td><td>0.000</td></tr>
<tr><td></td><td>GROWTH</td><td>0.397</td><td>0.486</td><td>0.816</td><td>0.426</td></tr>
<tr><td></td><td>INFLATION</td><td>−0.717</td><td>0.408</td><td>−1.756</td><td>0.097</td></tr>
<tr><td></td><td>GROWTH*INCUMBENT</td><td>0.340</td><td>0.494</td><td>0.688</td><td>0.501</td></tr>
</table>

[d]Dependent variable: VOTES.

effect (multiplicative) form [i.e., (5.9a)]. Indeed, we may consider both cases together in a single model and in doing so allow for both differences in intercepts and slopes. That is, we may estimate the following relationship:

$$Y_t = a + b_1 X_{1t} + b_2 X_{2t} + b_3 X_{3t} + b_4 (X_{1t} * X_{3t}) + e_t. \qquad (5.10)$$

The estimation of this model is left as an exercise for the reader (see Problem 5.4).[17]

Interaction effects can also be used to consider subcategories in models with several dummy variables. For example, we can return to our baseball player salary model shown in (5.7a). As we saw earlier, based on our regression analysis of our 32-player sample, there is little evidence that there is any difference in salaries in MLB between White, Black, and Non-Black Hispanic players, other things being equal. But what about the subcategory of players who are both Black *and* in the National League? Our dummy variable BLACK captures the effect of being a Black player (whether a National League player or an American League player), whereas the dummy LEAGUE controls only for league effects (whether a Black player or not). These two variables, however, do not allow us to consider the subcategory of Black, National Leaguers. It may be the case, for example, that Black players who are also in the National League have a salary differential as compared to other players. One method of capturing this subcategory is to create an interaction variable. This is done by simply multiplying the dummy variable BLACK times LEAGUE. That is, we create a new variable that is the product of X_{5i} and X_{7i}. Note that this new variable will be equal to 1 only if BLACK equals 1 (indicating a Black player) *and* LEAGUE equals 1 (indicating a National League player). In terms of our sample regression function, we can write it as follows:

$$Y_i = a + b_1 X_{1i} + b_2 X_{2i} + b_3 X_{3i} + b_4 X_{1i}^2 + b_8 (X_{5i} * X_{7i}) + e_i. \quad (5.11a)$$

Note that, for simplicity, we have left out the separate dummy variables for BLACK (X_{5i}), HISPANIC (X_{6i}), and NATIONAL LEAGUE (X_{7i}).[18] If a player is Black and plays in the National League, then X_{5i} and X_{7i} will be equal to 1, giving us

$$Y_i = a + b_1 X_{1i} + b_2 X_{2i} + b_3 X_{3i} + b_4 X_{1i}^2 + b_8 (1 * 1) + e_i, \quad (5.11b)$$

where b_8 captures the effect on salary for players who are *both* Black and in the National League as compared to all other players.

The Excel output provided in Table 5.5 shows the OLS estimate of (5.11a). As we can see, overall performance of the model has slightly improved as compared to the model without the interaction effect, judging by the adjusted R^2 (see Table 5.1). Using the regression results,

TABLE 5.5

Summary Output

Regression Statistics

Multiple R	0.915184763
R Square	0.83756315
Adjusted R Square	0.806325295
Standard Error	1.222656876
Observations	32

ANOVA

	df	SS	MS	F	Significance F
Regression	5	200.4082246	40.08164493	26.81244061	1.73513E − 09
Residual	26	38.86713572	1.494889835		
Total	31	239.2753604			

	Coefficients	Standard Error	t Statistic	P Value
Intercept	−35.57121376	21.76219099	−1.634541935	0.114196802
YEARS	1.149590555	0.184278289	6.238339643	1.3352E − 06
SLUGGING	0.017814216	0.003636274	4.899029935	4.38665E − 05
FIELDING	0.271202222	0.220013577	1.2326613	0.22873239 8
YEARS SQUARED	−0.047015538	0.00961414	−4.886551939	4.53358E − 05
BLACK*NL	0.590059163	0.562266852	1.049429042	0.303638047

we can write the predicted equation as (after rounding to three decimal places):

$$\widehat{SALARY}_i = -35.571 + 1.15(YEARS_i) + 0.018(SLUGGING_i)$$
$$+ 0.271(FIELDING_i) - 0.047(YEARS_i^2)$$
$$+ 0.59(BLACK_i * NATIONAL\ LEAGUE_i). \quad (5.11c)$$

As shown in Table 5.5, the coefficient to BLACK*NATIONAL LEAGUE is not statistically different from 0 at an acceptable level of significance. Nevertheless, it is instructive to interpret its coefficient. The estimated value for b_8 is 0.59, meaning that, other things being equal, Black players who are also in the National League earn, on average, about $0.59 million (or $590,000) more than other players. (Note that "other players" now includes White players, Non-Black Hispanic players, and Black players not in the National League.) Once again, however, these results are not reliable, given the large P value for the coefficient to BLACK*NL shown in Table 5.5.

As a final point, it should be noted that interactions between two continuous variables are possible. That is, we have thus far considered interactions between a dummy variable and a continuous variable [e.g., (5.9a)] and interactions between two dummy variables [e.g., (5.11a)]. We may also consider interactions between two continuous variables. For example, we can consider how player salaries in baseball differ for players who are not just good offensively, or good defensively, but players who are good at *both* offense and defense. We can represent this case with the following sample regression function:

$$Y_i = a + b_1 X_{1i} + b_2 X_{2i} + b_3 X_{3i} + b_4 X_{1i}^2 + b_9(X_{2i} * X_{3i}) + e_i. \quad (5.12a)$$

Note that this equation shows that the effects of, say, SLUGGING (X_{2i}), run through two parts of the equation. That is, suppose we consider the effects on a player's income of a one point increase in his slugging average (i.e., a one-unit increase in X_{2i}). As X_{2i} increases by one unit, salary is expected to increase by b_2, the coefficient to X_{2i}, *and* by $b_9 X_{3i}$. That is, the total effect of a one-unit increase in X_{2i} is equal to $b_2 + b_9 X_{3i}$. Thus, the effect of this one-unit increase in X_{2i} cannot be quantified until we know this player's value for fielding percentage, X_{3i}.[19] To estimate this new sample regression function, we first create a new variable, which is the product of SLUGGING and FIELDING. This new variable is then included on the list of independent variables along with the others shown in (5.12a). Table 5.6 presents the Excel output of this regression.

TABLE 5.6

Summary Output

Regression Statistics

Multiple R	0.938980391
R Square	0.881684175
Adjusted R Square	0.858931132
Standard Error	1.043479047
Observations	32

ANOVA

	df	SS	MS	F	Significance F
Regression	5	210.9652988	42.19305976	38.75016492	3.02463E − 11
Residual	26	28.31006155	1.088848521		
Total	31	239.2753604			

	Coefficients	Standard Error	t Statistic	P Value
Intercept	714.6367876	222.4182962	3.213030582	0.003488171
YEARS	1.212270124	0.156059717	7.767988747	3.05779E − 08
SLUGGING	−1.504581519	0.454517364	−3.31028391	0.002737238
FIELDING	−7.340405808	2.257785717	−3.251152557	0.003172756
YEARS SQUARED	−0.051711459	0.008224167	−6.28774463	1.17696E − 06
SLUGGING*FIELDING	0.015445346	0.00461361	3.347778073	0.00249163

Comparing the adjusted R^2's from this regression to that without the interaction effect (see Table 5.1), we see that this model considerably outperforms the latter. Furthermore, all of the estimated coefficients are statistically different from 0 at the 1% level of significance (indeed, the *largest P* value for the nonintercept coefficients is 0.003 for FIELDING). Thus, on the whole, our model is performing quite well. The predicted relationship is

$$\widehat{\text{SALARY}}_i = 714.637 + 1.212(\text{YEARS}_i) - 1.505(\text{SLUGGING}_i)$$
$$- 7.34(\text{FIELDING}_i) - 0.052(\text{YEARS}_i^2)$$
$$+ 0.015(\text{SLUGGING}_i * \text{FIELDING}_i). \qquad (5.12b)$$

Reviewing (5.12b), we can see that the coefficients to YEARS and YEARS SQUARED (i.e., YEARS^2) are nearly the same as they were in the regression without the interaction term for the slugging and fielding variables [see (5.3b)]. The coefficients for SLUGGING and FIELDING are quite different now and, in fact, may appear to be contrary to our theory. That is, both of these variables now have negative coefficients, which may seem to suggest that salaries tend to *decline* as offensive and defensive ability increase. This conclusion, however, would not be correct because we cannot assess the effects of these two independent variables without also considering their interaction. As noted earlier, the effect of SLUGGING has two components: the direct effect and the interaction effect. In this case, the coefficients shown in (5.12b), which are rounded to three decimal places, will not suffice as they will suffer from rounding errors. Thus, using the values for the estimated coefficients to SLUGGING and SLUGGING*FIELDING reported in Table 5.6 and carrying them out to five decimal places, we can see that the effect of a one-unit increase in SLUGGING is equal to $-1.50458 + 0.01544(\text{FIELDING})$. Thus, as stated previously, we cannot assess the effect of a unit change in SLUGGING until we know the value of FIELDING. For the purpose of evaluation, it is common to use the mean value of the interaction variable. In our sample of 32 MLB players, the mean value for FIELDING is 98.145. Thus, inserting this into our relation for the effects of a unit change in SLUGGING, we have

$$-1.50458 + 0.01544(98.145) = 0.01078.$$

Or, in other words, a one point increase in the slugging average of a player who has an average fielding percentage is expected to increase this player's salary by approximately $0.011 million, or $11,000, other things being equal. This salary increase may seem trivial, given that players in our sample have an average salary of $4.678 million. However, it should be noted that the preceding calculation was evaluated for a player with an average fielding percentage. If we consider a player who is an above-average fielder with, say, a fielding percentage of 99, then the expected addition to a player's salary for a one percentage point increase in his slugging average becomes

$$-1.50458 + 0.01544(99) = 0.02398,$$

or about $24,000. As we can see, the expected gain in salary is quite sensitive to changes in a player's fielding percentage. This result thus suggests that players who have good offensive *and* defensive skills are more highly valued than players who are skilled in just offense *or* defense; a result that is not unexpected. The effects of a one-unit increase in FIELDING would be calculated in a similar way and is left as an exercise for the reader (see Problem 5.5).

Time as an Independent Variable

When dealing with a time series data set, in some cases we may want to consider the effect time itself has on our dependent variable. That is, we may want to consider whether our dependent variable follows a **trend** (upward or downward) as time passes. As an example, we can consider a variation on our baseball example. Instead of considering the salary of our sample of 32 players for the year 1998, we can consider the average wage of *all* MLB players *over* the years. Many people believe that MLB salaries, on average, have been increasing (i.e., following an upward trend) from year to year. This hypothesis can be tested by collecting data on average baseball salaries over the years and then *creating* a **time index,** which is simply a variable that keeps track of time periods. For example, we can consider the following linear model:

$$Y_t = \alpha + \beta t + u_t, \tag{5.13}$$

where Y_t is the average salary paid to a player in year t and this is shown as a linear function of t itself.[20] In this case, β represents the increase in the average player's salary from one year to another. The corresponding sample regression function for (5.13) would be

$$Y_t = a + bt + e_t. \tag{5.14a}$$

The OLS method can be used to estimate the equation shown in (5.14) and it will be BLUE so long as the regression model assumptions are satisfied (see Chapter 2).[21]

To illustrate this task of fitting a trend to a time series of data, consider the data on average baseball salaries shown in Table A.2 of Appendix A. The data show average player salaries (in dollars) for the years 1969 through 1997. The time index we will use, which will serve as our independent variable, will be the column for YEAR shown in the table. Using Excel to calculate the OLS regression for (5.14), we obtain the results shown in Table 5.7. The R^2 of 0.853 tells us that about 85.3% of the variation in salaries over these years can be explained by our model. Using the values for the intercept and the coefficient to YEAR (rounded to two decimal places), we can write the estimated sample regression function as

$$\widehat{Y}_t = -93916191.92 + 47576.54t. \tag{5.14b}$$

As we can see, the intercept has no reasonable interpretation. Technically, it tells us that, if $t = 0$, then the expected salary is about $-\$93.9$ million. This unrealistic result is due (in part) to the fact that our data for time do not actually go back to YEAR equal to 0. The coefficient to our time index tells us, on average, how salaries have increased from one year to the next. The value for b, whose P value shows that the coefficient is statistically different from 0 at a very small level of significance (i.e., a very high level of confidence), suggests that salaries increased, on average, by about \$47,576.54 each year. Given (5.14b), we can now use it to **forecast** salaries for future years. For example, suppose we wanted to forecast average baseball salaries for the year 2001. All we need to do is to plug in the value 2001 into (5.14b) for t. Doing so, we obtain

$$\widehat{Y}_{2001} = -93916191.92 + 47576.54(2001) = 1284464.6. \tag{5.14c}$$

TABLE 5.7

Summary Output

Regression Statistics

Multiple R	0.923844358
R Square	0.853488397
Adjusted R Square	0.848062041
Standard Error	170921.3206
Observations	29

ANOVA

	df	SS	MS	F	Significance F
Regression	1	$4.59496E+12$	$4.59496E+12$	157.2857455	$8.98204E-13$
Residual	27	$7.88781E+11$	29214097852		
Total	28	$5.38374E+12$			

	Coefficients	Standard Error	t Statistic	P Value
Intercept	−93916191.92	7522718.321	−12.48434248	$9.98199E-13$
YEAR	47576.54532	3793.571036	12.54136139	$8.98204E-13$

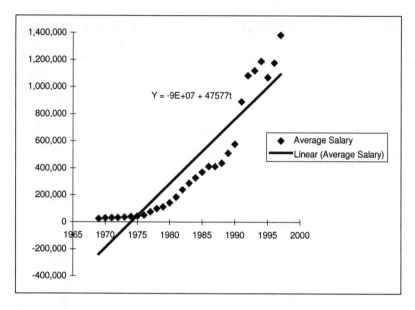

Figure 5.4.

Thus, forecasted average salaries for the year 2001 are a little more than $1.28 million.

The model shown in (5.14a) is a simple linear trend. In many cases, however, *nonlinear* trends may better reflect the behavior of a time series. Indeed, many believe that baseball salaries have not only increased over the years, but they have increased at an *increasing rate*. If this is true, then a linear trend is not the appropriate functional form for our forecasting equation and we will need to estimate a nonlinear trend.[22] A simple way of determining whether a linear or a nonlinear trend is appropriate is to plot the data and check them visually. Figure 5.4 shows the plot of average baseball salaries over the years 1969 to 1997. Also included in this plot is the estimated trend line (i.e., the sample regression function) from (5.14b).[23]

We can clearly see in Figure 5.4 that the linear trend is a poor representation of the data. In fact, average salaries do appear to increase at an increasing rate over time; thus, a nonlinear trend is warranted. To fit a nonlinear trend to the data, we can use a quadratic functional form for our sample regression function such as

$$Y_t = a + b_1 t + b_2 t^2 + e_t. \tag{5.15a}$$

To estimate the sample regression function shown in (5.15a), we first create another independent variable, which is simply the squared value of the time index, YEAR. We can then use Excel to calculate the OLS values for a, b_1, and b_2 in (5.15a). Doing so yields the results shown in Table 5.8. The output shown in Table 5.8 indicates that, overall, our model is performing quite well. The R^2 of approximately 0.97 tells us that 97% of the variation in salaries over the years is explained by our model.[24] The very large F statistic indicates that our regression is significant at a very small significance (high confidence) level. As for the estimated coefficients, all three have very small P values, indicating that they are significantly different from 0 at a very small significance (high confidence) level. The intercept is positive, which is more realistic than the negative value that we found for our simple linear trend, but it still has no useful interpretation because YEARS and YEARS SQUARED never have 0 values in our data set. The coefficient for b_1 is negative and for b_2 it is positive. These together will produce a J shaped nonlinear trend with the trend declining in the beginning, but increasing at an increasing rate later.[25] Using the calculated values for b_1 and b_2 from Table 5.8, we can write the estimated nonlinear a trend (i.e., the sample regression function) as

$$\widehat{Y}_t = 9143047086 - 9268739.85t + 2349.05t^2. \qquad (5.15b)$$

Judging by their P values, both b_1 and b_2 are statistically significant from 0 at a very small level of significance (i.e., high level of confidence), thus supporting the hypothesis that, over time, salaries have increased at an increasing rate. Figure 5.5 provides a plot of the data along with the nonlinear trend.[26] It is evident that this model provides a better fit than the simple linear trend.

Using (5.15b), we can again forecast expected average salaries for the year 2001 by simply plugging in this value for t. We have the following:

$$\widehat{Y}_{2001} = 9143047086 - 9268739.85(2001) + 2349.05(2001)^2 = 1897195.2. \qquad (5.15c)$$

Thus, the result in (5.15c) forecasts an average salary for baseball players of approximately \$1.9 million for the 2001 season.[27]

TABLE 5.8

Summary Output

Regression Statistics

Multiple R	0.984676147
R Square	0.969587115
Adjusted R Square	0.967247663
Standard Error	79356.8181
Observations	29

ANOVA

	df	SS	MS	F	Significance F
Regression	2	$5.22001E+12$	$2.61E+12$	414.4504078	$1.90436E-20$
Residual	26	$1.63735E+11$	6297504579		
Total	28	$5.38374E+12$			

	Coefficients	Standard Error	t Statistic	P Value
Intercept	9143047086	92717016.8	9.86129839	$2.8339E-10$
YEAR	−9268739.855	935132.2117	−9.911689213	$2.54934E-10$
YEAR SQUARED	2349.045991	235.7868263	9.962583696	$2.29206E-10$

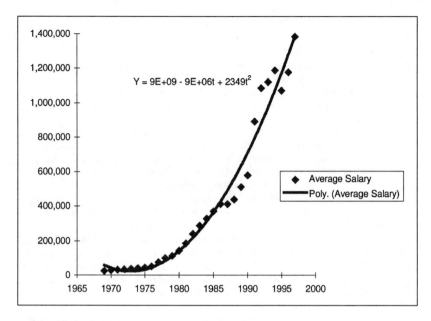

Figure 5.5.

SUMMING UP

We have seen in this chapter that a variety of types and forms of independent variables are possible. Nonlinear independent variables can be incorporated into our regression models to allow for increasing or decreasing effects of independent variables on the dependent variable. In cases where one or more independent variables are not continuous, but are instead categorical, we can create dummy variables to control for categorical effects. Finally, we saw that time can also be used as a variable when working with time series data. By fitting a trend (either linear or nonlinear) to data, we can understand how the dependent variable tends to change across time. Once obtained, this trend can be used to perform elementary forecasting.[28]

▼ Notes

1. This is analogous to the "learning curve." In the beginning of a new activity, a participant learns very quickly about performing the activity. Later, additional experience teaches the participant little about performing the activity.

2. If both β_1 and β_2 are positive, then the function will be increasing but at an increasing rate.

3. This is easily accomplished in both SPSS and Excel. In SPSS, the Transform item on the main menu bar is chosen, then Compute is chosen. At the next dialog box, the Target Variable is the name to be given to the new variable (e.g., YEARSSQ, for YEARS squared). In the Numeric Expression box is where the equation is typed to create the new variable. For example, we would type YEARS**2 to create the square of YEARS, and then select OK. As for Excel, one would simply create a new column of data headed with, say, YEARSSQ. In the first entry below this label, we could type the equation, $= (\text{cell address})^2$, where "cell address" would be the location of the first element below YEARS, and hit Enter. This would create the square of the first observation's value for YEARS and put below the label YEARSSQ. Once this is done, the copy and paste commands can be used to copy the newly created equation to all the cells below it for the entire sample.

4. See note 20 of Chapter 4. Note that although including a variable and its square as independent variables does not violate the no perfect multicollinearity assumption, it may lead to problems of high multicollinearity. This issue is discussed briefly in Chapter 6.

5. Readers familiar with differential calculus may note that these calculations can also be made by first taking the *partial derivative* of (5.3) with respect to the variable X_{1i} and then plugging in the values 2 and 5 into the derivative.

6. Reciprocal relationships can also be incorporated. That is, we can include an X variable and its reciprocal, $1/X$.

7. This can be done, for example, by taking the natural logs of the dependent variables, the independent variables, or both (see Problem 5.1). For more details on various functional forms and transformations, see Gujarati (1995).

8. In contrast to this scenario, George Bush was the incumbent-party candidate in the 1988 election, but he was not the incumbent.

9. Dummy variables are sometimes called "indicator," "categorical," or "binary" variables.

10. By excluding a relevant variable from our model, we would be violating CLRM assumption 5 regarding model specification error (see Chapter 2).

11. It is possible to use other numbers to represent categories, but using 1 and 0 makes the interpretation of our results easier and more intuitive.

12. This test is somewhat trivial in light of the fact that the regression without INCUMBENT was statistically significant and by adding this dummy variable our model is strengthened.

13. The reason for this advantage may be due to the fact that the president typically has much more exposure to voters through the media than other candidates and he can use this exposure to his benefit.

14. For a more extensive discussion on the use of dummy variables, see Hardy (1993). It is also possible that the dependent (Y) variable could be a dummy variable. This topic, however, is beyond the scope of this book and the reader is referred to Aldrich and Nelson (1984).

15. In the event that a player is both Black and Hispanic, he is classified as Black in this example.

16. It is often the case that models that include interaction effects frequently suffer from extreme multicollinearity, (a subject we take up in Chapter 6), which may explain the lack of significance of these two variables.

17. Models that include both dummy variables *and* interaction effects using the same dummy variables are even more likely to suffer from extreme multi-collinearity than models that just contain one or the other.

18. We could include the separate dummy variables BLACK, HISPANIC, and NATIONAL LEAGUE. We have seen in a previous regression, however, that the coefficients to these variables were not statistically different from 0. In addition, as was noted in the case of our presidential voting model (see note 17), inclusion of an interaction variable along with its associated components (i.e., BLACK*NATIONAL LEAGUE, BLACK, and NATIONAL LEAGUE) often leads to severe multicollinearity problems.

19. Those readers comfortable with differential calculus can determine the affects of a change in X_{2i} on Y_i by taking the partial derivative of (5.12a) with respect to X_{2i}. Doing so, we have $\partial Y_i / \partial X_{2i} = b_2 + b_9 X_{3i}$. The value of this partial derivative is an increasing function of X_{3i}, showing that players with strong offensive *and* defensive skills are more highly valued than players with just strong offensive skills, other things being equal.

20. It should be made clear that we are now considering the average salary paid to *all* MLB players in a particular year.

21. In this type of analysis, the regression model assumption of "no autocorrelation" (see Chapter 2, CLRM assumption 3) will frequently be violated and the OLS results will not be strictly valid. In this case, other methods of estimating the regression model are necessary. This issue is touched on in Chapter 6.

22. If we estimate a linear model when the data clearly follow a nonlinear trend, then we are committing a "specification error," the results of which may invalidate our estimated sample regression function. See Gujarati (1995) for more details.

23. Excel's chart function includes an option that allows a trend line to be plotted to a set of data points in the chart.

24. It is not uncommon to obtain a very high R^2 in trend analyses of this type.

25. If both b_1 and b_2 are positive, then the function will increase at an increasing rate, and it will not decline anywhere. Here, given that b_1 is negative while b_2 is positive, we have a small decline in the beginning of the trend, but it later increases at an increasing rate thereafter.

26. Note that the equation for the nonlinear trend embedded in Figure 5.5 is the same as that shown in (5.15b), except for some rounding and the use of scientific notation.

27. Not bad work if you can get it!

28. Forecasting methods can become quite complex. For details, see, for example, Diebold (1998).

PROBLEMS

5.1 In (5.2), the *level* of MLB player salaries is our dependent variable. The coefficients to the variables included in our model (i.e., b_1, b_2, b_3, and b_4) show how a player's salary increases for a one-unit increase in the related X variable. Economic theories of labor, however, suggest that wage earners tend be rewarded with *relative* increases in salary (i.e., percentage increases) and not dollar increases. One way of estimating such a relationship is to use the *natural log* of salaries, instead of actual salaries, for the dependent variable. Furthermore, by carrying out this natural-log transformation, the nonlinear relationship we had before (as shown in Figure 5.1) may be "straightened out," leaving us with a linear relationship. That is, consider the following sample regression function (not including the dummy variables for race):

$$\ln Y_i = a + b_1 X_{1i} + b_2 X_{2i} + b_3 X_{3i} + e_i,$$

where the natural log of salaries ($\ln Y_i$) is shown as a simple linear function of years of MLB experience (X_{1i}), slugging average (X_{2i}), and fielding percentage (X_{3i}). In this case, the OLS-produced estimates of b_1, b_2, and b_3 show the *relative* change in salary for a one-unit change in the associated X's. Multiplying the coefficient by 100 thus gives the estimated *percentage* change in salary for a one-unit change in the X's. For example, if the OLS estimate for b_1 is equal to 0.10, then $b_1 * 100 = 10$. Thus, if YEARS increases by 1, then SALARY is expected to increase by 10%. Using the data in Table A.1, do the following:

a. Use SPSS or Excel (or an equivalent program) to calculate the natural log of SALARY and estimate the previous regression using OLS.

b. Interpret the estimated coefficients to the X variables.

5.2   Two other factors that may affect the abortion rate in a state is whether or not the state provides public funds that can be used for abortions and whether or not there are laws, such as required parental consent, that may restrict access to abortion services. To control for these potential factors, two dummy variables are included in Table A.4: FUNDS, which equals 1 if a state provides public funds for abortions, 0 otherwise; and LAWS, if a state enforces laws that may restrict access to abortions.

a. Reestimate (4.11) adding FUNDS and LAWS as independent variables and interpret the values of the coefficients for these two variables.

b. Does this model outperform the one without FUNDS and LAWS? Explain.

c. Are the coefficients to FUNDS and LAWS statistically different from 0 at the 5% significance (95% confidence) level?

5.3   Time variables can be used in a regression along with other variables. Using the data in Table A.3, reestimate (5.5) adding a time index. (Hint: Create a time index based on the election year. For example, the election year 1916 can be $t = 1$, election year 1920 can be $t = 2$, etc.) Compare your results to those in Table 5.2. Is the coefficient to the time index statistically different from 0 at the 10% significance (90% confidence) level?

5.4   Models can be estimated that allow for differences in intercepts (using a dummy variable) *and* slopes (using an interaction effect). Such a model is shown in (5.10). Estimate this model using OLS and interpret the results. (Note that some of the estimated coefficients will turn out not to be statistically different from 0. As a matter of practice, interpret these coefficients, but with the understanding that their values are not reliable results.)

5.5   Using the results shown in Table 5.6, and given that the mean value for SLUGGING is 474.488, determine the expected effect on a player's salary for a one-unit increase in FIELDING.

5.6   Returning to Problem 4.4, which models wage as a function of
education and experience, we can build on this model by considering various categorical variables. Specifically, we can consider the effects of gender, marital status and race. Reviewing
Table A.6, we see the columns headed FEMALE, MARRIED, and
BLKHISP. These three dummy variables indicate whether the particular individual is female, married, or either Black or Hispanic; in
each case, a 1 indicates "yes," and a 0 indicates "no."

   a. Given these definitions for FEMALE, MARRIED, and BLKHISP,
      what is the base category?

   b. Estimate the sample regression model with WAGE as the de-
      pendent variable and EDUC, EXPER, FEMALE, MARRIED,
      and BLKHISP as the independent variables and interpret the
      estimated coefficients. Which coefficients are statistically dif-
      ferent from 0 at the 10% level of significance?

SOME COMMON PROBLEMS IN REGRESSION ANALYSIS

In Chapter 2, the regression model assumptions were introduced with a brief explanation of their significance. Later, in Chapter 4, we added the assumption of "no perfect multicollinearity." It was noted that if these assumptions are met, then OLS produces the best possible estimation of our model, or OLS is BLUE. This is clearly a powerful result.

It is often the case, however, that one or more of the regression model assumptions are not satisfied. In this case, OLS is typically *not* the best method for estimating our sample regression function. The purpose of this chapter is to briefly discuss some of the most common problems that occur in regression analysis and to map out directions for further study. The goal is to provide the reader with an intuitive understanding of the problem at hand and how it might be solved. The application of these proposed solutions is an advanced topic and the reader is pointed to more advanced books on these subjects.

INTRODUCTION

The Problem of High Multicollinearity

Recall, from Chapter 4, that perfect multicollinearity is the case when one of our independent variables has a perfect linear relationship to one or more of the other independent variables in our model. If this occurs, then we have perfectly redundant information between these X variables and we are unable to use the least-squares method to estimate such a model.[1] Our solution to this problem is simple: Drop one of the redundant variables.

A related problem that is somewhat more difficult to solve is the case of *high multicollinearity*. This occurs when one of the independent variables has nearly a perfect linear relationship to one or more of the other X variables. The intuition as to why this may be problematic is not too difficult to grasp. Suppose we have one X variable that is not perfectly linearly related, but very closely related to one or more other X variables. If this is the case, then there is a considerable overlap in the information that these variables contain about the behavior of our dependent variable, Y. Because these X variables share much of the same explanatory power over Y, they rob each other of significance. This being the case, each variable on its own may appear to be statistically insignificant, yet together they may be highly significant. Put another way, if two or more X variables look very similar in their information about the behavior of Y, then the OLS procedure may not be able to distinguish the *unique* explanatory ability of one X variable from the others. The solution to such a problem may seem simple, drop one of these *highly* (but not *perfectly*) redundant variables. The problem with this solution is that, in dropping one of the variables, we may be compromising the theory behind our model. *It is important to understand that multicollinearity does not signal a problem with our theory, rather it is the inability of our data to clearly distinguish the separate yet subtle effects of two independent variables.*

As an illustration of the problem of high multicollinearity, we can return to our example of presidential elections. In Chapter 4, we estimated a model showing the percentage of two-party votes received by the incumbent-party candidate as a function of economic growth and inflation. This model is reproduced as follows:

$$Y_t = a + b_1 X_{1t} + b_2 X_{2t} + e_t, \tag{6.1}$$

where X_{1t} is our measure of economic growth and X_{2t} is our measure of inflation. Later, in Chapter 5, we considered a voting model that added the interaction effect between growth and incumbent. This equation is reproduced as follows:

$$Y_t = a + b_1 X_{1t} + b_2 X_{2t} + b_4 (X_{1t} * X_{3t}) + e_t, \tag{6.2}$$

TABLE 6.1

(A)
Coefficients[a]

Model		Unstandardized Coefficients		t	Sig.
		B	Std. Error		
1	(Constant)	54.339	1.964	27.668	0.000
	GROWTH	0.700	0.203	3.450	0.003
	INFLATION	−0.553	0.326	−1.694	0.108

[a]Dependent variable: VOTES.

(B)
Coefficients[b]

Model		Unstandardized Coefficients		t	Sig.
		B	Std. Error		
1	(Constant)	55.148	2.315	23.827	0.000
	GROWTH	0.397	0.486	0.816	0.426
	INFLATION	−0.717	0.408	−1.756	0.097
	GROWTH*INCUMBENT	0.340	0.494	0.688	0.501

[b]Dependent variable: VOTES.

where $(X_{1t} * X_{3t})$ is our interaction effect for growth and incumbency (X_{3t}). Recall that when (6.1) was estimated using OLS we found that the estimated coefficient to growth (b_1) was statistically different from 0 at better than the 1% level of significance. When (6.2) was estimated, the coefficient to growth (b_1) was no longer statistically significant at a reasonable level of significance. For purposes of comparison, both regression results are shown in Table 6.1.

Thus, the addition of the interaction effect has led to the apparent loss in significance of a variable that previously was statistically significant. In this event, we should suspect a problem with high multicollinearity.[2] If we consider what kind of relationship we may have among the X variables, we would immediately suspect that perhaps growth and the interaction effect are closely related. One method of detecting whether we have a problem of high multicollinearity is to calculate the **correlation coefficient** between the two variables. The correlation coefficient is a measure showing the degree to which any two variables are linearly related. A correlation coefficient of 1 implies that two variables are perfectly, positively related. A correlation coefficient of −1 implies a perfect negative relationship. A value of 0 implies that the two variables are not linearly related.[3] In cases where two independent variables have a correlation coefficient of 0.8 or greater (in absolute terms), then this may signal a high multicollinearity problem. In our voting model, the correlation coefficient between growth (X_{1t}) and the interaction term ($X_{1t} * X_{3t}$) is 0.858; thus, high multicollinearity is apparently why the coefficient to growth (b_1) is not statistically different from 0 in the model that includes both of these independent variables.

Given evidence that high multicollinearity is present in our data, we may ask what it means for our OLS estimation method and what we can do to alleviate the problem. With regard to the first issue, the most important point to remember is that, *in the presence of high multicollinearity, OLS is still BLUE*. That is, high multicollinearity is a nuisance, but it doesn't destroy any of the desirable properties of the ordinary least-squares method.

As for remedies to this problem, several methods are possible; however, they often bring problems of their own.[4] The best solution to the problem of high multicollinearity is to try to increase the sample size of our data set. As noted earlier, multicollinearity does not signal a problem with our model. Rather, it reflects the inability of the OLS method to distinguish the subtle, separate effects of our X variables on Y, given our data set. In the case of our model shown in (6.2), it may be that if we had a larger sample size, then the added information contained in this larger sample would enable OLS to distinguish the separate, subtle effects of our interaction term from the effects of growth on its own. Of course, it may also be true that our interaction term is truly unimportant in explaining our dependent variable. In any event, given that presidential elections are relatively few and far between, it seems unlikely that we would be able to increase our sample size. Thus, in this case given that the inclusion of the interaction term did not improve the overall performance of our regression model (the adjusted R^2, in fact, dropped when the interaction term was added), we may drop the interaction term from the model.

Other potential cases of multicollinearity problems are not so clear-cut. For example, recall our MLB model. In Chapter 4, we estimated a sample regression function showing player salaries to be a function of years of MLB experience, slugging average, and fielding percentage [see (4.5)]. The results, provided in Table 4.1, showed that all three of these independent variables were statistically different from 0 at better than the 1% level of significance. Later, in Chapter 5, this model was reestimated including a new variable, years of MLB experience squared [see (5.2)]. Table 5.1 presented the results of the OLS estimation of this second model and we found that, when YEARS SQUARED was included in our model, the adjusted R^2 increased, implying a better fit. We also found, however, that the coefficient to the variable FIELDING was no longer statistically different from 0 at an acceptable level of significance (e.g., 10% or smaller). Thus, the addition of a new variable has led to the apparent loss in significance of a variable that previously was statistically significant. Again, in this event, we should suspect a problem with high multicollinearity. We can proceed as we did in our earlier example and calculate the correlation coefficient between FIELDING and YEARS SQUARED. Doing so, we find a value of 0.037, which is clearly a long ways from 0.8, our benchmark for indicating high correlation. Correlation coefficients between pairs of variables, however, are not capable of capturing more complex relationships between independent variables. For example, FIELDING may not be linearly related to YEARS SQUARED, but it may be linearly related to a *combination* of YEARS, YEARS SQUARED, and SLUGGING. This seems reasonable, especially with regard to YEARS and YEARS SQUARED because the longer a player plays in the major leagues, the better fielder he may become. This (potential) relationship among the X variables can be represented, in fact, as a regression function:

$$\text{FIELDING} = \alpha + \beta_1(\text{SLUGGING}_i) + \beta_1(\text{YEARS}_i) + \beta_2(\text{YEARS}_i^2). \quad (6.3)$$

If there is indeed a significant relationship between FIELDING and the other independent variables shown in (6.3), then this would suggest that there is an overlap among these variables in our original regression function shown in (5.2). Using our sample of data, we can estimate the sample regression function for (6.3) and check to see if this **auxiliary regression** produces a reasonably high R^2. If it does, then this would be evidence of high multicollinearity between FIELDING and the other three independent variables. Table 6.2 shows the results

TABLE 6.2

Summary Output

Regression Statistics

Multiple R	0.507275209
R Square	0.257328138
Adjusted R Square	0.177756153
Standard Error	1.112829464
Observations	32

ANOVA

	df	SS	MS	F	Significance F
Regression	3	12.01449637	4.004832124	3.233903709	0.037185044
Residual	28	34.67490363	1.238389415		
Total	31	46.6894			

	Coefficients	Standard Error	t Statistic	P Value
Intercept	98.03733093	1.495332781	65.56221609	$3.39026E-32$
YEARS	0.44344006	0.143350252	3.093430247	0.004452338
SLUGGING	−0.00324174	0.003176366	−1.020581479	0.316190358
YEARS SQUARED	−0.021984621	0.007642058	−2.876793392	0.007601234

of this estimated auxiliary regression function. Reviewing Table 6.2, we can see that the R^2 is approximately 0.257, indicating a rather poor fit. Thus, there is no strong evidence of high multicollinearity between FIELDING and the remaining independent variables. This does not necessarily mean, however, that multicollinearity problems are not present. It simply means that the auxiliary regression did not clearly detect multicollinearity. Other methods of detecting multicollinearity may, in fact, provide evidence of its presence.[5]

Nonconstant Error Variance

When working with cross-section data sets, a problem that frequently comes up is nonconstant error variance. This problem relates to the second regression model assumption that was discussed in Chapter 2 and has to do with the spread, or "variance," of the population error terms, u_i, around the population regression function. Recall that we assumed that the error variance is constant throughout the regression line. In this case, the errors are said to be "homoscedastic," meaning "constant variance." This was shown graphically in Figure 2.6 and is shown again in Figure 6.1(A).

If the error variance is not constant throughout the population regression line, thus violating our second regression model assumption, then the errors are said to be "heteroscedastic," meaning "nonconstant variance." A graphic example would be where our observations tend to "funnel out" as we move out along the population regression function. An example is provided in Figure 6.1(B). Alternatively, the observations may "funnel in" as shown in Figure 6.1(C). To understand the intuition as to why nonconstant error variance may occur, we can return to our example on abortion rates in the United States.

Recall our model as shown in (4.11) where the abortion rate in state i is hypothesized to be a function of the state i's religious makeup (RELIGION), the average price of an abortion (PRICE), the average income (INCOME), and the frequency of antiabortion activities (as measured by PICKETING). We can focus our attention on the variable INCOME. If we think of abortion services as simply a service that people can consume, we can consider the different purchasing behavior of people with high versus low income.[6] People with high income have greater freedom in choosing how to spend (or not spend) their money

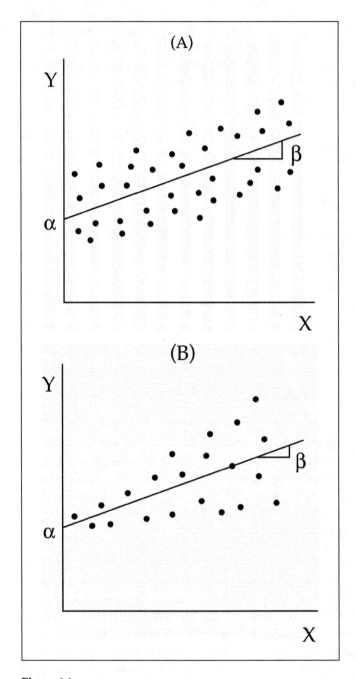

Figure 6.1.

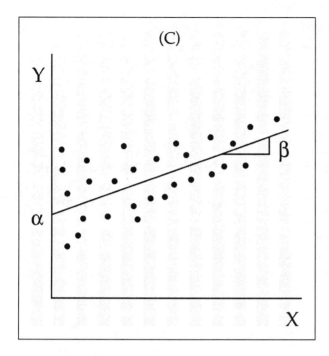

Figure 6.1. (Continued).

and as such we may witness greater variation in the way individuals consume. Thus, although abortion rates tend to be larger in higher-income states, on average (see Table 4.3), there may also be a greater spread around the average, reflecting the fact that in some high-income states individuals decide not to purchase abortion services, whereas in other high-income states individuals choose to purchase this service. In low-income states, this story is reversed. That is, because individuals in these states have lower incomes, on average, they have less freedom on how to spend their income and, this being the case, spending behavior would tend to have less variation. Graphically, we can represent this scenario in Figure 6.2 by considering, for the sake of this discussion, just two groups of states: high-income and low-income states.

We can let X_{HIGH} represent the value of income for high-income states and X_{LOW} represent the value of income for low-income states. The variable Y represents the abortion rate. The fact that our line slopes upward reflects the hypothesis that, *on average*, the abortion rate increases as income increases. Considering high-income states, the heavy dot on the line may represent the average abortion rate for those states, with the actual values for individual states' abortion rates falling above or below the group's average. Similarly for low-income states, the heavy dot on the line may represent this group's average abortion rate, with individual states' values falling above and below the group average. The key aspect of Figure 6.2 is that the individual observations for low-income states are more closely speckled around their group's mean, whereas individual observations from the high-income group are more broadly speckled around their mean. This reflects the point made earlier that those with larger incomes have greater choice of how to spend them and thus we may witness greater variation in their purchases of abortion services, with the reverse being true for those with low incomes. Recalling that the vertical distance from an individual observation to the regression line is equal to the error term, then we can see the sum of the squared errors for the low-income group is smaller

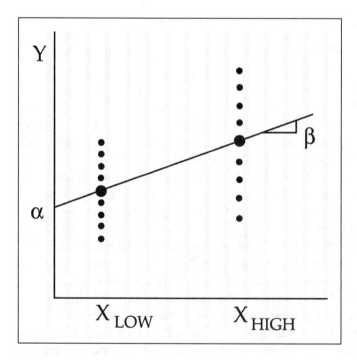

Figure 6.2.

than the sum of the squared errors of the high-income group. Or, using more formal language, the error variance increases (and is hence "nonconstant") as incomes increase. This kind of behavior of the error variance may also be true at incomes between X_{HIGH} and X_{LOW}, thus producing the "funneling out" shape as shown in Figure 6.1. In fact, Figure 6.3 plots our actual data on abortion rates and income levels for the 50 states. The graph of the data has a generally upward sloping relationship as our hypothesis predicts. The dots also appear to slightly funnel out as income gets larger, indicating potential nonconstant error variance.

Another graphical method that is commonly used to detect heteroscedasticity is first to perform a least-squares regression and save the residuals from the regression.[7] Then we create a plot with the squared residuals on the vertical axis and the variable we suspect may be the source of the heteroscedasticity, INCOME, on the horizontal axis.[8] If there is no pattern in the errors (i.e., if the errors are homoscedastic),

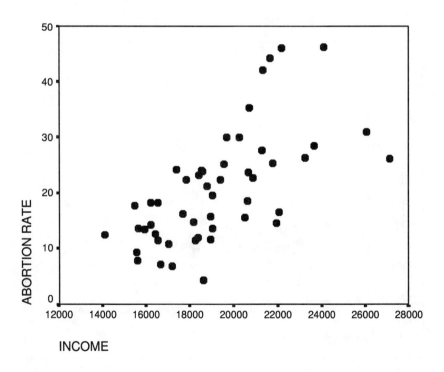

Figure 6.3.

then the plot should show a fairly uniform horizontal band of dots. If it is true that the error variance increases as INCOME increases, then this plot should show the squared errors getting larger as INCOME increases. Figure 6.4 shows the plot of the graph. As shown in Figure 6.4, the squared residuals get larger as INCOME increases, indicating heteroscedasticity.[9]

Given that heteroscedasticity is present, what does this mean for our OLS method? Unfortunately, it means that OLS is no longer BLUE. It is still a linear unbiased estimator, but it is no longer the best. Recall that "best" meant that the method produced sample regression estimates of the intercept term, a, and the b's with the smallest variance. Or, in other words, with the greatest accuracy. Furthermore, the OLS-produced standard errors of the intercept and coefficients to the X variables are not valid. This means that the associated t statistics and P values for the a and b's in our model are not valid and we cannot legitimately perform hypothesis tests.

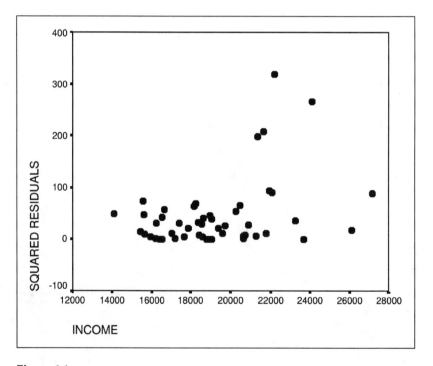

Figure 6.4.

Given detection of heteroscedasticity and the problems it causes for OLS, we can now consider, briefly, a possible remedy. There are, in fact, a number of proposed solutions to this problem, but the most common one is that of **weighted least squares** (WLS). Although the details of this method are somewhat complicated, the intuition behind this method is not too difficult to follow. Consider again Figure 6.2, where we have two income groups shown: states with low income (X_{LOW}) and states with high income (X_{HIGH}). Recall our general goal, to draw a sample of data from the population and use it to find a sample regression function that best fits the data. As shown in Figure 6.2, sample observations from the X_{LOW} group have a smaller variance as they are more tightly centered around the group's mean (the heavy dot), as compared to sample observations that are drawn from the X_{HIGH} group. Thus, on average, sample observations drawn from X_{LOW} are more reliable predictors of their group mean (because they tend to be closer to their heavy dot) than are observations from X_{HIGH} as predictors of their group mean. Weighted least squares takes note of this difference and when fitting a line to the data WLS assigns greater weight to observations from the more reliable, low-variance group (X_{LOW} in this example) and less weight to the less reliable, high-variance group (X_{HIGH} in this case). Estimating a sample regression function in this way can produce results that are, in fact, BLUE.

As noted, details of performing a WLS estimation are somewhat complicated and the reader is referred to other sources for a complete discussion.[10] Some regression programs, such as SPSS, do include the WLS method and the user must provide some information about how to reweight observations to take into account nonconstant error variance.[11] Table 6.3 provides the SPSS results of a WLS estimation that was performed for our model on state abortion rates. In performing this WLS estimation, the data are reweighted by dividing all observations by the square root of INCOME (i.e., the variable SQRTINC).[12] Notice in Table 6.2 that the WLS estimates for the coefficients to RELIGION, PRICE, INCOME, and PICKETING have the same sign and, except for RELIGION, are similar in size to the OLS results shown in Table 4.3. Also, as before, the coefficients to PRICE, INCOME, and PICKETING are significantly different from 0 at the 5% significance (95% confidence) level or better. Thus, our results have not changed substantially, except now they are valid.[13]

TABLE 6.3

		Model Summary		
Model	R	R Square	Adjusted R Square	Std. Error of the Estimate
1	0.731[a]	0.535	0.493	85.3654409

[a]Predictors: (Constant), PICKETING WITH CONTACT, PRICE, RELIGION, INCOME.

ANOVA[b,c]

Model		Sum of Squares	df	Mean Square	F	Sig.
1	Regression	376856.701	4	94214.175	12.929	0.000[d]
	Residual	327926.633	45	7287.259		
	Total	704783.334	49			

[b]Dependent variable: ABORTION RATE.

[c]Weighted least squares regression—weighted by SQRTINC.

[d]Predictors: (Constant), PICKETING WITH CONTACT, PRICE, RELIGION, INCOME.

Coefficients[e,f]

Model		Unstandardized Coefficients		t	Sig.
		B	Std. Error		
1	(Constant)	−5.032	9.227	−0.545	0.588
	RELIGION	$4.526E-03$	0.086	0.053	0.958
	PRICE	$-4.56E-02$	0.022	−2.042	0.047
	INCOME	$2.363E-03$	0.000	6.144	0.000
	PICKETING	−11.558	4.128	−2.800	0.008

[e]Dependent variable: ABORTION RATE.

[f]Weighted least squares regression—weighted by SQRTINC.

Autocorrelated Errors

A common problem with time series data is that of autocorrelation. Recall that the third regression model assumption of Chapter 2 was that the error from one observation is not related to the error of another. If, however, the errors are related, then we face the problem of autocorrelation. As with nonconstant error variance, the details of this problem are complex, but the intuition behind this problem is fairly easy to grasp.

Essentially, what the assumption of no autocorrelation means is that there is no systematic relationship among the population errors (i.e., the u_t's). That is, if we know one period's error, it should tell us nothing about the next period's error. If this is not the case, and there is some systematic relationship among the errors, then our OLS method is no longer BLUE and other methods are needed.[14]

There are several reasons why there may be a systematic relationship, or pattern in the errors. First, in time series data, it may be the case that any error in one period can carry over to subsequent periods because it may take several periods before the error is fully absorbed. For example, we may consider a model that estimates the price of a company's stock. Suppose the company suffers an unexpected financial setback. This may be reflected in a fall in the company's stock price that our model would not have predicted (because it was unexpected) and our error for this period would be negative (because the actual value of the stock price would be below our predicted value, other things being equal). This fall in stock prices, however, may be spread over several periods and the next period's error may also be negative. In other words, sometimes errors tend to have momentum such that an error in one period tends to carry over to subsequent periods.

Another common reason for patterns in the errors is model misspecification. For example, recall Figure 5.4, where a line is fitted to data that are apparently nonlinearly related. In this case, we see that the first group of observations all fall above the linear regression function and thus they would all have positive errors. The next group of observations all fall below the line and thus would all have negative errors. The third group falls above the line and they would all have positive errors. Thus, we clearly have a pattern in the errors in this case, as we would observe strings of positive errors and strings of negative errors.

Concerning the *types* of patterns in the errors, there are two basic patterns commonly observed. The first is the kind suggested in the stock price example, namely, that one period's error *tends* to have the same sign as the previous period's error. This is called "positive autocorrelation," and if we graphed the errors across time they would generate a picture like that shown in Figure 6.5(A). The second basic type of autocorrelation is when one period's error tends to have the opposite sign of the previous period's error. This type of pattern, called "negative autocorrelation," is shown in Figure 6.5(B), which has a sawtooth look to it, indicating continual sign flips.

There are several methods that can be used to detect patterns in the errors.[15] A simple approach is to examine the errors visually from an OLS regression and see if there is, in fact, any distinguishable pattern. As an example, we can return to our model from Chapter 5 that looks at average player salaries in MLB over the years 1969 to 1997. Using our nonlinear model as shown in (5.15a) (so as to avoid any model misspecification bias), we can rerun this regression, save the errors, and analyze them. Table 6.4 provides a printout of the residuals.

If we concentrate on the signs of the residuals, we can see an apparent pattern. The first three residuals are negative, the next four positive, then five negative residuals, followed by fifteen positive ones, and ending with two negative residuals. The fact that we have long strings of negative and positive errors is a sign that we may have autocorrelation problems. If the errors were truly random and not related to each other, we would not expect to see such strings.[16] Another way to look for patterns in the residuals is to plot them across time and look for apparent patterns in the plot. Figure 6.6 shows such a plot for the residuals given in Table 6.4. As shown in Figure 6.6, the residuals follow a kind of wave pattern across time, which is similar to the positive autocorrelation graph shown in Figure 6.5(A).

Given detection of autocorrelation, we can consider remedies to this problem. Details of common remedies are, in fact, quite complex and the reader is encouraged to read more about this issue in more advanced books.[17] The basic idea behind these remedies is the following. Given that the errors have a pattern to them, this suggests that there is some systematic information in the errors that is going unused in our OLS estimation method. One technique commonly used, called the **Cochrane–Orcutt procedure**,[18] tries to discover the systematic component in the errors and then use this information to refine our estimates

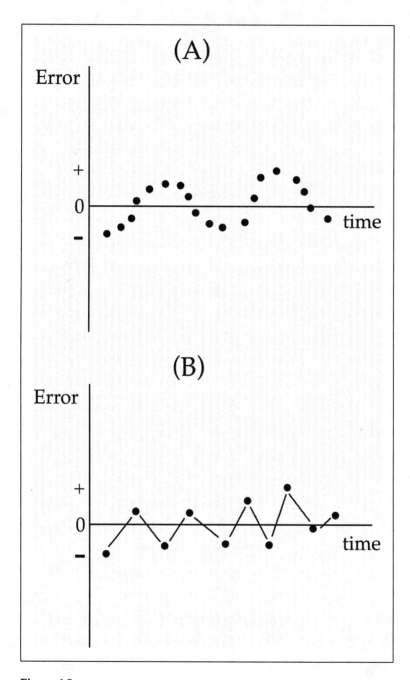

Figure 6.5.

TABLE 6.4

	Residual Output	
Observation	Predicted Average Salary	Residuals
1	1390148.88	−6570.88
2	1279148.09	−102181.09
3	1172845.39	−101816.39
4	1071240.79	117438.21
5	974334.28	145919.72
6	882125.86	202282.14
7	794615.53	96572.47
8	711803.30	−132873.30
9	633689.16	−121605.16
10	560273.10	−121544.10
11	491555.15	−79101.15
12	427535.28	−15015.28
13	368213.50	3357.50
14	313589.82	15818.18
15	263664.23	25529.77
16	218436.73	23060.27
17	177907.32	7743.68
18	142076.01	1679.99
19	110942.78	2615.22
20	84507.65	15368.35
21	62770.61	13295.39
22	45731.67	5769.33
23	33390.81	11285.19
24	25748.05	15090.95
25	22803.38	13762.62
26	24556.80	9535.20
27	31008.31	534.69
28	42157.91	−12854.91
29	58005.61	−33096.61

of the coefficients. In essence, we internalize the systematic information in the errors and, in so doing, produce a better fit for our data. The
details regarding this technique are somewhat complicated and are beyond the scope of this book. As a means of comparison with our results
in (5.15b) (which does not correct for autocorrelation), (6.4) provides the
estimated regression function for (5.15b) after applying the Cochrane–

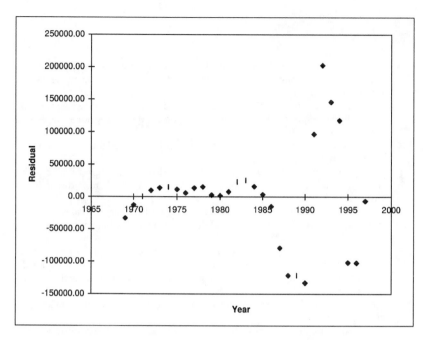

Figure 6.6.

Orcutt procedure to correct for autocorrelated errors:

$$\widehat{Y}_t = 8.94714(E+9) - 9.07147(E+6)t + 2299.39t^2,$$

t statistic:	4.361	-4.380	4.399
P value:	0.00001	0.00001	0.0001

$$(6.4)$$

Comparing these results (which use scientific notation) to those in (5.15b), we see that the estimated values of a, b_1, and b_2 are very similar in size. As before, the small P values indicate that the coefficients to time (t) and the square of time (t^2) are statistically different from 0 at a very small level of significance (i.e., very high level of confidence). The difference here is that these coefficient estimates and tests of significance are now valid.[19]

▼ Notes

1. Recall our example in Chapter 4 of considering the two independent variables, birth date and age. These are perfectly redundant because, knowing one, we can determine the other.

2. This is one of several signs of near-perfect multicollinearity. There are others, such as unexpected signs on variables. For more details, see, for example, Gujarati (1995).

3. Both Excel and SPSS are capable of calculating correlation coefficients for pairs of variables.

4. For example, we can use a method called "ridge regression," but this method brings with it other problems. See Judge et al. (1985) for a discussion of this potential remedy.

5. In fact, an alternative method of detecting multicollinearity, called the "condition index" [see Gujarati (1995)], did produce evidence of high multicollinearity in our estimation of (5.2).

6. To reiterate what was stated in previously Chapter 4, thinking of abortion services as simply another service that consumers can purchase is done for the purpose of facilitating our discussion and is not intended to trivialize this socially important subject.

7. Both Excel and SPSS have the option of saving the residuals from a regression.

8. Squaring the errors will often make subtle patterns in the errors more obvious and easier to detect.

9. There are many other tests for heteroscedasticity. See Gujarati (1995) for more details.

10. See, for example, Gujarati (1995) or Greene (2000).

11. Excel does not have a WLS routine provided as part of the program. An experienced user of Excel, however, could perform a WLS estimation with proper programming.

12. See Gujarati (1995) for a discussion on the choice of weighting schemes in performing the WLS.

13. Using more formal language, our WLS results are BLUE *asymptotically* (i.e., for large samples).

14. If we use OLS when autocorrelation is present, then the estimated variance is likely to be biased and as such this makes our t statistics and F statistics invalid.

15. The most common one is the Durbin–Watson test. See Gujarati (1995) for details about this test.

16. If the errors are truly random and unrelated, we would expect the signs of the errors to be more random than what we observe in this example. That is, in this case, we see too few sign flips. Note that another pattern in the residuals could be *too many* sign flips (i.e., errors that systematically oscillate between positive and negative values). There is, in fact, a formal test of autocorrelation, called the "runs test," that focuses on the signs of errors. See Geary (1970) for details.

17. For example, Gujarati (1995).

18. See Cochrane and Orcutt (1949).

19. To be more precise, the Cochrane–Orcutt procedure will produce estimates that are BLUE and t statistics that are valid *asymptotically* (i.e., in large samples).

PROBLEMS

6.1 One factor that may be important in determining an MLB player's salary is the number of games that he has played over his career. Players who play in a large number of games over their career are likely more important to their team, and hence should command a higher salary, than players who play in fewer games, all else being equal. Suppose a variable called GAMES, equal to the number of career games a player has appeared in, is added to our model shown in (5.2). How might the inclusion of this variable cause a problem with high multicollinearity? Explain.

6.2 Recall our model shown in Problem 4.5 for estimating MSPR prices for SUVs as a function of horsepower and engine size. Suppose we add to this model the following three independent variables:

CARGO = cargo space measured in cubic feet
MPG = miles per gallon, averaged between city and highway
CYLINDERS = number of cylinders in engine

The OLS estimate of this model produces the following estimated sample regression function:

$$\widehat{MSRP}_i = -16162.23 - 54.91(CARGO_i) + 215.78(MPG_i)$$
$$- 2532.53(SIZE_i) + 8366.93(CYLINDERS_i)$$
$$+ 28.61(HORSEPOWER_i),$$

with $R^2 = 0.693$ and adjusted $R^2 = 0.646$. Notice that CARGO and SIZE have negative coefficients, implying that cars with greater cargo area and larger engines are expected to have *lower* prices. This result seems counterintuitive. Additionally, none of the estimated coefficients, with the exception of the one for CYLINDERS, is statistically different from 0 at the 10% level of significance. Thus, given the relatively high R^2, the counterintuitive signs, and the few significant coefficients, we seem to have a classic case of high multicollinearity. As a means of convincing us of this diagnosis, run auxiliary regressions for the five independent variables and report the estimated R^2 values.

6.3   One of the many ways of testing for nonconstant error variance is the **Glejser test** [see Glejser (1969)]. The intuition behind this test is easy to follow. Recall that with nonconstant error variance, or heteroscedasticity, the error variance is not uniform along the sample regression line. For example, Figures 6.3 and 6.4 both indicate for our abortion example that the absolute size of the error tends to increase as INCOME increases. The Glejser test simply tries to show, using regression analysis, whether the absolute size of the error is a function of one (or more) of the independent variables. That is, for the abortion rate example, we may have

$$|e_i| = a + b_1 \text{INCOME}_i + w_i, \qquad (6.5)$$

where $|e_i|$ is the absolute value of the error, shown to be a function of INCOME. The term w_i in (6.5) represents an error term, indicating that the relationship between $|e_i|$ and INCOME (if any) would likely be imperfect. The hypothetical relationship in (6.5) can be estimated using OLS by first estimating the equation shown in (4.11) and saving the residuals from this regression. Next, the absolute value of the residuals can be computed and then used as the dependent variable in (6.5) with INCOME as the independent variable. If the resulting OLS regression for (6.5) shows that the coefficient to INCOME is statistically different from 0, then this provides evidence that heteroscedasticity is present. Using the data in Table A.4, test for heteroscedasticity using the Glejser method and explain your results.

6.4   The presidential elections model shown in (5.4a) is estimated with a time series of data and, as such, is a candidate for autocorrelation problems. Estimate the model shown in (5.4a), plot the residuals from this regression, and visually inspect them for signs of autocorrelation. What are your conclusions?

7

WHERE TO GO FROM HERE

O ne of the goals of this book was to show the reader how the tools of regression analysis can be put to work on a wide variety of research topics and subject areas. Regression analysis is regularly applied to research in economics, political science, sociology, education, the physical sciences, as well as other fields. It is a tool that can be applied in many creative ways. It is also a tool, however, that is often misused. Indeed, much of the criticism aimed at regression analysis can be traced to an irresponsible use of the methods discussed in this book.

INTRODUCTION

Key Points to Bear in Mind

There are several things that a practitioner of regression analysis needs to bear in mind:

- Regression analysis is based on probability and statistical principles and, as such, the *theory* of regression analysis is impeccable. The *application* of regression analysis, on the other hand, is in the hands of the researcher. If it is applied responsibly, it can be a powerful research tool. If it is applied irresponsibly, it can be very misleading.
- In most cases, regression analysis is incapable of *proving* anything. This is because research is typically carried out with *samples* of data. Sample results can be convincing and can *support* hypotheses, but proof beyond any doubt is beyond the ability of any sample of data.

- Computer programs that perform regression analysis simply obey the commands of the user. These programs are not capable of building theoretically sound models. This is the (often difficult) task of the researcher. A well-constructed model can provide us with valuable results. Applying regression analysis to a poorly constructed models can be misleading.
- The examples used in this text were carried out with small-sized samples of data. This was done purposefully so that the reader could easily input these data by hand into a program capable of performing OLS regressions. In general, samples should be as large as feasibly and reasonably possible. Larger samples generally provide more reliable estimates of population regression functions.
- The language of regression analysis is important. This book has at times used informal language in order to provide the reader with intuition. In the application of regression analysis, the researcher should endeavor to use formal language when appropriate. For example, when discussing a random variable, observations do not literally "bounce around" their mean, which was a phrase that was used in Chapter 4; rather, they have *variance*.

Other Topics in Regression Analysis

The methods used in this book are the most basic to regression analysis. There are quite literally dozens of other topics under regression analysis that can be pursued. Many of these topics have to do with the nature of the dependent variable. A few examples of specialized topics follow:

- *Dummy dependent variables.* We have already discussed dummy independent variables in Chapter 5. In some cases, however, it is the dependent variable that may be categorical. For example, we may wish to study the voting behavior of U.S. senators on a particular political issue. In this case, the dependent variable, the behavior of which we wish to explain, would be the observed vote cast by a senator. For example, we may code a senator's vote as 1 equals a "yes" vote on the issue, 0 equals a "no" vote. We could then construct a model in which we specify independent variables that we believe explain the likelihood (i.e., the *probability*) that a senator will vote, say, in favor of the issue. In this case, given that the dependent variable is a dummy, it is inappropriate to use our OLS method to estimate such a model. There are several appropriate estimation methods such as the *logit* and *probit* estimation methods. See DeMaris (1992), Menard (1995), and Liao (1994) for a discussion of these types of models.

- *Pooled data sets.* In Chapter 1, it was noted that there are three basic data set types: cross-section, time series, and pooled data. In this book, we have worked with cross-section data (e.g., the baseball player and the abortion data sets) and time series data (e.g., the presidential election data and the yearly average MLB salary data). We have not, however, worked with pooled data (which is sometimes referred to as *longitudinal* or *panel* data). An example of pooled data would be, say, following the salaries of a cross section of baseball players from one year to the next. Or, perhaps, a data set that tracks the abortion data of the 50 U.S. states from one year to the next. In both of these examples, we have a combination of cross-section data and time series data. These kinds of data sets are quite common and may require special techniques in order to make full use of both the cross-section information in the data and the time series information. For techniques on handling pooled data, see Menard (1991) and Hsiao (1990).

- *Simultaneous equations methods.* In some cases, to understand the behavior of a particular dependent variable, it must be considered in the context of a *system* of equations. That is, thus far our sample regression models have considered the behavior of a dependent variable in a single equation. In many cases, a single dependent variable may be a component of linked equations. For example, consider our abortion model. We have considered the demand for abortion as being a function of several independent variables, including the price of an abortion. Students of economics will recall, however, that the market price and the quantity of a good or service that is produced and consumed is determined by *both* supply and demand for the good or service. Thus, to fully understand demand for abortion, we should also consider supply simultaneously.[1] A discussion of special techniques for handling such *simultaneous equations* models can be found in Gujarati (1995) or Greene (2000).

Go Forward and Regress!

We have come a long way from the simple two-variable model that was introduced in Chapter 2. The multiple regression model we developed in Chapter 4, together with the discussion in Chapter 5 on various independent variable types, has left us with a powerful analytical tool that can be used to explore a great variety of topics. As this book comes to an end, the reader is encouraged to make use of the tools we have discussed. However, as the title of this book suggests, we have covered the basics of regression analysis and there is a great deal more to learn. Thus, the reader is also encouraged to continue to read and learn more about advanced regression analysis techniques such as those we have discussed previously.[2] The discussion in this introductory-level book has provided the basic ideas and intuition of regression analysis and the reader should now be well equipped to pursue more advanced topics in this field.

———•◆•———

▼ Notes

1. See Kahane (2000) and Medoff (1988) for a simultaneous equations approach to understanding the market for abortion services.

2. There are a number of good books available that cover a broad range of topics from simple to advanced methods in regression analysis. These include the ones cited throughout this book such as Gujarati (1995) and Greene (2000). The bibliographies in these books list numerous other sources.

APPENDIX A
DATA SETS USED IN EXAMPLES

DATA SETS

TABLE A.1   MLB Data

Player	1998 Salary	Team	League AL = 0, NL = 1	Career 97 MLB Years	Career 97 Slugging	Career 1997 Fielding %	Black 1 = yes, 0 = no	Hispanic 1 = yes, 0 = no
Abreu, Bob	0.18	PHI	1	1	372	97.92	1	0
Anderson, Brady	5.442	BAL	0	11	434	98.78	0	0
Bagwell, Jeff	7.945	HOU	1	7	535	99.25	0	0
Bonds, Barry	8.917	SF	1	12	551	98.44	1	0
Boone, Bret	2.800	CIN	1	5	398	98.88	0	0
Bordick, Mike	3.583	BAL	0	7	327	97.89	0	0
Damon, Johnny	0.400	KC	0	2	387	98.63	0	0
Galarraga, Andres	8.000	ATL	1	12	494	99.17	0	1
Griffey, Ken Jr.	7.979	SEA	0	9	560	98.62	1	0
Grissom, Marquis	5.000	MIL	1	8	413	98.75	1	0
Guerrero, Vladimir	0.230	MON	1	1	483	93.37	1	0
Gwynn, Tony	4.000	SD	1	16	454	98.63	1	0
Hamilton, Darryl	2.750	COL	1	8	381	99.37	0	0
Higginson, Bobby	2.425	DET	0	3	499	97.33	0	0
Justice, David	6.500	CLE	0	8	511	97.72	1	0
Karros, Eric	4.500	LA	1	6	455	99.23	0	0
McGwire, Mark	8.333	STL	1	10	554	99.21	0	0
Molitor, Paul	4.250	MIN	0	19	451	96.97	0	0
O'Neill, Paul	5.450	NYY	0	11	472	98.98	0	0
Piazza, Mike	8.000	LA	1	5	575	98.87	0	0
Ripken, Cal Jr.	6.300	BAL	0	17	450	97.8	0	0
Rodriguez, Alex	2.113	SEA	0	3	539	96.52	1	0

TABLE A.1 (Continued)

Player	1998 Salary	Team	League AL = 0, NL = 1	Career 97 MLB Years	Career 97 Slugging	Career 1997 Fielding %	Black 1 = yes, 0 = no	Hispanic 1 = yes, 0 = no
Rodriguez, Ivan	6.600	TEX	0	7	437	98.88	0	1
Salmon, Tim	5.000	ANA	0	5	527	97.59	0	0
Sosa, Sammy	8.000	CHC	1	8	468	96.87	1	0
Stairs, Matt	0.325	OAK	0	2	511	97.05	0	0
Thomas, Frank	7.000	CWS	0	8	601	99.06	1	0
Vaughn, Greg	5.250	SD	1	9	452	98.4	1	0
Vaughn, Mo	6.600	BOS	0	7	532	98.79	1	0
Ward, Turner	0.750	PIT	1	3	434	99	0	0
Williams, Matt	4.800	ARI	1	11	497	96.25	0	0
Zaun, Greg	0.280	FLA	1	1	441	98.42	0	0

1998 Salary = reported salary, in millions of dollars, the player earned for the 1998 MLB season.

League = 1 if the player was playing in the National League, 0 if in the American League.

Career 97 MLB Years = number of seasons through 1997 the player has played at least 130 at bats.

Career 97 Slugging = slugging average through 1997, calculated as the ratio (number of bases reached)/(number of at bats)*1000.

Career 97 Fielding % = fielding percentage through 1997, calculated as the ratio (assists + putouts)/(assists + putouts + errors)*100

Black = 1 if a player is Black, 0 otherwise.

Hispanic = 1 if a player is Non-Black Hispanic, 0 otherwise.

DATA SETS

TABLE A.2   MLB Yearly Average Salary Data

Year	Average Salary
1997	1,383,578
1996	1,176,967
1995	1,071,029
1994	1,188,679
1993	1,120,254
1992	1,084,408
1991	891,188
1990	578,930
1989	512,084
1988	438,729
1987	412,454
1986	412,520
1985	371,571
1984	329,408
1983	289,194
1982	241,497
1981	185,651
1980	143,756
1979	113,558
1978	99,876
1977	76,066
1976	51,501
1975	44,676
1974	40,839
1973	36,566
1972	34,092
1971	31,543
1970	29,303
1969	24,909

SOURCE: Sean Lahman's Web page: www.baseball1.com (reprinted with permission).

TABLE A.3 Presidential Election Data

Year	Votes	Incumbent	Growth	Inflation
1916	51.68	1	2.229	4.252
1920	36.12	0	−11.463	16.535
1924	58.24	1	−3.872	5.161
1928	58.82	0	4.623	0.183
1932	40.84	1	−15.574	6.657
1936	62.46	1	12.625	3.387
1940	55	1	2.42	0.553
1944	53.77	1	2.91	6.432
1948	52.37	1	3.105	10.369
1952	44.6	0	0.91	2.256
1956	57.76	1	−1.479	2.132
1960	49.91	0	0.02	2.299
1964	61.34	1	4.95	1.201
1968	49.6	0	4.712	3.16
1972	61.79	1	5.716	4.762
1976	48.95	0	3.411	7.604
1980	44.7	1	−3.512	7.947
1984	59.17	1	5.722	5.296
1988	53.9	0	2.174	3.392
1992	46.55	1	1.478	3.834
1996	54.6	1	2	2.3
Mean:	52.48			

Year = election year.

Votes = percentage of the two-party vote received by the incumbent-party candidate.

Incumbent = 1 if the candidate is the incumbent, = 0 otherwise.

Growth = growth rate, in percent, of real GDP over the three quarters prior to the election.

Inflation = inflation rate, in percent, over the 15 quarters prior to the election.

TABLE A.4   Abortion Data

State	ABORTION	RELIGION	PRICE	LAWS	FUNDS	EDUC	INCOME	PICKET
Alabama	18.20	36.40	272.00	1.00	0.00	66.90	16522	89
Alaska	16.50	18.10	461.00	0.00	1.00	86.60	22067	0
Arizona	24.10	29.30	249.00	0.00	0.00	78.70	17401	55
Arkansas	13.50	30.00	248.00	1.00	0.00	66.30	15635	33
California	42.10	28.10	293.00	0.00	1.00	76.20	21348	36
Colorado	23.60	21.20	309.00	0.00	0.00	84.40	20666	43
Connecticut	26.20	43.40	374.00	0.00	1.00	79.20	27150	60
Delaware	35.20	19.80	247.00	0.00	0.00	77.50	20724	0
Florida	30.00	22.60	271.00	1.00	0.00	74.40	19711	37
Georgia	24.00	28.60	319.00	0.00	0.00	70.90	18549	50
Hawaii	46.00	26.70	422.00	0.00	1.00	80.10	22200	0
Idaho	7.20	36.80	303.00	0.00	0.00	79.70	16649	50
Illinois	25.40	37.10	272.00	0.00	0.00	76.20	21774	47
Indiana	12.00	16.50	288.00	1.00	0.00	75.60	18366	67
Iowa	11.40	29.20	280.00	0.00	0.00	80.10	18275	50
Kansas	22.40	21.30	340.00	1.00	0.00	81.30	19387	67
Kentucky	11.40	36.70	320.00	0.00	0.00	64.60	16528	75
Louisiana	13.40	50.90	228.00	1.00	0.00	68.30	15931	60
Maine	14.70	22.40	328.00	0.00	0.00	78.80	18163	0
Maryland	26.40	22.80	264.00	0.00	0.00	78.40	23268	50
Massachusetts	28.40	50.10	330.00	1.00	1.00	80.00	23676	70
Michigan	25.20	28.00	352.00	1.00	0.00	76.80	19586	28
Minnesota	15.60	44.60	270.00	1.00	0.00	82.40	20503	67

TABLE A.4 (Continued)

State	ABORTION	RELIGION	PRICE	LAWS	FUNDS	EDUC	INCOME	PICKET
Mississippi	12.40	38.00	256.00	0.00	0.00	64.30	14082	100
Missouri	11.60	32.20	348.00	1.00	0.00	66.90	18970	50
Montana	18.20	26.90	329.00	0.00	0.00	81.00	16227	50
Nebraska	15.70	30.90	279.00	1.00	0.00	81.80	18974	100
Nevada	44.20	23.50	275.00	0.00	0.00	78.80	21648	33
New Hampshire	14.60	27.80	372.00	0.00	0.00	82.20	21933	50
New Jersey	31.00	42.70	316.00	0.00	1.00	76.70	26091	64
New Mexico	17.70	44.70	332.00	0.00	0.00	75.10	15458	20
New York	46.20	41.80	338.00	0.00	1.00	74.80	24095	60
North Carolina	22.40	25.90	291.00	0.00	1.00	70.00	17863	54
North Dakota	10.70	56.30	370.00	1.00	0.00	76.70	17048	100
Ohio	19.50	24.70	298.00	1.00	0.00	75.70	19040	60
Oklahoma	12.50	36.30	281.00	0.00	0.00	74.60	16420	75
Oregon	23.90	15.80	248.00	0.00	1.00	81.50	18605	50
Pennsylvania	18.60	37.00	296.00	0.00	0.00	74.70	20642	82
Rhode Island	30.00	63.90	322.00	1.00	0.00	72.00	20276	50
South Carolina	14.20	30.30	292.00	1.00	0.00	68.30	16212	57
South Dakota	6.80	38.90	400.00	0.00	0.00	66.30	17198	100
Tennessee	16.20	30.90	300.00	0.00	0.00	67.10	17674	43
Texas	23.10	41.80	257.00	0.00	0.00	72.10	18437	56
Utah	9.30	76.70	298.00	1.00	0.00	85.10	15573	0
Vermont	21.20	26.60	276.00	0.00	1.00	80.80	18792	50
Virginia	22.70	19.90	267.00	0.00	0.00	75.20	20883	38

DATA SETS

TABLE A.4 (Continued)

State	ABORTION	RELIGION	PRICE	LAWS	FUNDS	EDUC	INCOME	PICKET
Washington	27.70	17.80	270.00	0.00	1.00	83.80	21289	24
West Virginia	7.70	9.80	251.00	0.00	1.00	66.00	15598	50
Wisconsin	13.60	41.60	276.00	1.00	0.00	78.60	19038	67
Wyoming	4.30	29.30	378.00	1.00	0.00	83.00	18631	100
Mean:	20.58	32.65	305.12	0.36	0.24	75.93	19215.52	52

ABORTION = number of abortions per thousand women aged 15 to 44 in 1992.

RELIGION = percentage of a state's population who is either Catholic, Southern Baptist Evangelist, or Mormon.

PRICE = average charged in 1993 in nonhospital facilities for an abortion at 10 weeks with local anesthesia (weighted by the number of abortions performed in 1992).

LAWS = variable that takes the value of 1 if a state enforces a law that restricts a minor's access to abortion, 0 otherwise.

FUNDING = variable that takes the value of 1 if state funds are available for use to pay for an abortion under most circumstances, 0 otherwise.

EDUC = percentage of a state's population who is 25 years or older with a high school degree (or equivalent), 1990.

INCOME = disposable income per capita, 1992.

PICKET = percentage of respondents who reported experiencing picketing with physical contact or blocking of patients.

NOTE: These data are available for downloading on the Internet at www.sbeusers.csuhayward.edu/~lkahane/index.html.

TABLE A.5 Sport Utility Vehicle Data

Vehicle	MSRP	CARGO	MPG	ENGINE SIZE	CYLINDERS	HORSEPOWER
BMW X5 4.4i AWD 0062, 4d	49970	54.4	15	4.4	8	282
CADILLAC ESCALADE, 4d	46925	118.2	14	5.7	8	255
CHEVROLET BLAZER ls sport, 4d	25245	74.1	19	4.3	6	190
CHEVROLET TAHOE, 4wd, 4d, 2000	39594	118.2	14	5.7	8	255
CHEVROLET TRACKER, 4d, 4wd, hard top, auto	16725	44.7	24	2	4	127
DODGE DURANGO, 4wd, 4d	28870	88	14	5.9	8	245
FORD EXPEDITION, u16, xlt 4d, 4wd	33405	118.3	16	4.6	8	215
FORD EXPLORER XLS utility, 4d, 4wd, auto	25935	79.8	18	4	6	160
GMC JIMMY SLE Sport utility, 4d, 4wd	29245	74.1	18	4.3	6	190
GMC YUKON, 4wd, 4d	35305	108.2	16	4.8	8	275
HONDA CR-V LX, 4wd, auto	20290	67.2	23.5	2	4	146
HONDA PASSPORT LX, 4wd, auto	27540	81.1	18	3.2	6	205
INFINITY QX4	36075	85	16.5	3.3	6	170
ISUZU AMIGO, 4wd, hard top, 2d	20990	62.5	17.5	3.2	6	205
ISUZU RODEO, 4wd, 4d, auto	24935	81.1	18	3.2	6	205
ISUZU TROOPER L44, 4wd, 4d, auto	29445	90.2	17	3.5	6	215
ISUZU VEHICROSS, 2d, auto	31045	50.4	17	3.5	6	215
JEEP CHEROKEE SE, 4wd, 4d	19885	69	19	2.5	4	125
JEEP GRAND CHEROKEE Laredo, 4wd, 4d	29425	72.3	18.5	4	6	195
JEEP WRANGLER, 4wd, 2d, soft	15020	55.2	19	2.5	4	120
KIA SPORTAGE, 4wd, 4d auto	18045	55.4	20.5	2	4	130
LAND ROVER DISCOVERY Series II, SD, 4wd	33975	70.3	15	4	8	188
LAND ROVER RANGE ROVER 4.0, 4wd	59625	58	15	4	8	188
LEXUS LX470	60600	90.4	14.5	4.7	8	230
LEXUS RX300, 4wd	35100	75	21.5	3	6	220

TABLE A.5 (Continued)

Vehicle	MSRP	CARGO	MPG	ENGINE SIZE	CYLINDERS	HORSEPOWER
LINCOLN NAVIGATOR, 4wd	47100	117.6	14.5	5.4	8	300
MERCEDES-BENZ M-CLASS 320	35945	85.4	18	3.2	6	215
MERCURY MOUNTAINEER	30135	80.2	18	4	6	210
MITSUBISHI MONTERO	32302	67.1	17.5	3.5	6	197
MITSUBISHI MONTERO SPORT LS, 4wd	27302	79.3	18.5	3	6	165
NISSAN PATHFINDER XE, 4wd, auto	28919	85	17	3.3	6	170
NISSAN XTERRA, 4wd, auto	22269	65.6	15.5	3.3	6	170
OLDSMOBILE BRAVADA	32173	74.1	18	4.3	6	190
SUBARU FORESTER, L AWD, auto	20590	64.6	24.5	2.5	4	165
SUZUKI VITARA JX, 4wd,4d, auto	18549	44.7	24	2	4	127
TOYOTA 4RUNNER SR5, 4wd, auto	28838	79.8	18	3.4	6	183
TOYOTA LAND CRUISER	52208	90.8	14.5	4.7	8	230
TOYOTA RAV4, 4wd, 4d, auto	19828	57.9	24	2	4	125

MSRP = manufacturers suggested retail price for the base model.

CARGO = cargo space measured in cubic feet.

MPG = miles per gallon, averaged between city and highway.

ENGINE SIZE = size of engine in liters.

CYLINDERS = number of cylinders in engine.

HORSEPOWER = engine horse power.

SOURCE: Kelly Blue Book Web page: www.kbb.com.

TABLE A.6   Wage Data

WAGE	EDUCATION	FEMALE	MARRIED	EXPERIENCE	BLACK/ HISPANIC
11	8	0	1	42	0
20.4	17	0	0	3	0
9.1	13	0	0	16	1
13.75	14	0	1	21	0
24.98	16	0	1	18	0
7.7	13	1	0	8	0
10	10	1	1	25	0
3.4	8	1	0	49	0
7.38	14	1	0	15	1
22	14	0	1	15	0
6.67	16	1	0	10	0
12	14	1	1	10	0
9.25	12	0	0	19	0
5	12	0	0	4	1
4.13	12	1	1	4	0
3.65	11	0	0	16	0
7	12	1	1	10	0
22.83	18	1	0	37	0
5	12	1	1	14	0
22.5	18	0	1	14	0
4.5	12	1	0	3	0
6	12	0	1	9	0
6.75	10	0	1	13	0
8.75	12	0	0	9	0
3.35	16	1	1	14	1
6	13	0	0	31	0
7.3	12	0	1	37	0
4.25	12	1	1	20	0
6.25	18	0	1	14	0
5.75	12	1	1	15	0
3.35	11	0	0	3	1
7.45	12	1	0	25	0

TABLE A.6   (Continued)

WAGE	EDUCATION	FEMALE	MARRIED	EXPERIENCE	BLACK/ HISPANIC
3.43	12	0	0	2	1
4.35	8	1	1	37	1
5.85	18	0	1	12	0
8.5	12	1	0	16	0
7.5	12	1	1	27	1
3.35	12	1	1	10	1
4.22	8	1	1	39	0
5	12	1	1	39	0
9.6	12	1	1	14	0
12.57	12	0	1	12	0
12.5	13	1	1	16	0
8.89	16	1	1	22	0
6.25	9	0	0	30	0

WAGE = hourly wage in dollars (1991 figures).

EDUCATION = years of education.

FEMALE = 1 if the individual is a female, 0 if male.

MARRIED = 1 if the individual is married, 0 otherwise.

EXPERIENCE = years of labor market experience.

BLACK/HISPANIC = 1 if the individual is Black or Hispanic, 0 otherwise.

APPENDIX B
INSTRUCTIONS FOR USING
EXCEL AND SPSS

The purpose of this appendix is to provide the reader with some basic instructions on how to use the programs Excel and SPSS to perform the regression analyses carried out in the text. This appendix assumes that the reader is somewhat familiar with using spreadsheet programs like Excel. For a more complete discussion on how to use Excel and SPSS, the reader is referred to the program manuals.[1]

For each program, we consider how to do the following:

- Input data, including inputting data by hand, reading data in from a data file, and saving data as a program file. (The instructions for reading data in from data files are given for the on-line data sets used in the examples in this book.[2] Data obtained from elsewhere may require different steps for reading into the programs, depending on their format.)
- Transform variables into new variables.
- Perform a simple least-squares regression, including how to save residuals from a regression.

Using Excel

Microsoft Excel is a spreadsheet program that is capable of doing simple least-squares regression analysis. Regression analysis, however, is not the primary intention of the program and, as such, performing OLS with Excel can at times be somewhat cumbersome as compared to other programs (e.g., SPSS).

Inputting Data

By Hand

Data can be entered into Excel by typing in values directly. Data should be entered as columns, where each column contains all observations on a single variable. In the first row of each column, there should be a variable name followed immediately by the associated data. Columns may contain nonnumerical entries, such as player names. Table B.1 shows an (abbreviated) example of how the baseball data set would look when properly entered into Excel. (Variable definitions are as they were given in the text.)

Reading Data from a Data File

Rather than typing data in by hand, it is more convenient to read data from data files, if such files are available. Indeed, by reading data from a data file, there is less of a chance of incorrectly inputting data values as there is when doing so by hand.[3] Data files can be of a variety of formats. The most general format is the ASCII or "text" file. The on-line data available for use in this book's examples are text files and the following steps describe how to read in these data. To read a data text file into Excel, it is assumed that the data have already been downloaded onto a computer disk and saved as a text file. Having done so, the following steps are carried out for reading the text data into Excel:

1. Initiate Excel. Then, from the main tool bar menu, choose File, then Open, and choose the appropriate location for Look in. Provide the appropriate File Name: (e.g., bball.txt), and under Files of type: select Text Files, then select Open.
2. The next screen shows Excel's Text Import Wizard. Under Original data type, choose Delimited (the default), then select Next.
3. At the next screen, select Tab for Delimiters (the default) and for Text Qualifier, select (double quotation mark, again the default). Select Next.
4. At this screen, set the Data Format to General (the default) and then select Finish.

The data should now appear as they do in Table B.1. Before continuing, *you should save the data set* by going to File on the main menu bar. Select Save As .... Choose a location and file name and change the Save

TABLE B.1

PLAYER	SALARY	TEAM	NL	YEARS	SLUGGING	FIELDING	BLACK	HISPANIC
Abreu, Bob	0.18	PHI	1	1	372	97.92	1	0
Anderson, Brady	5.442	BAL	0	11	434	98.78	0	0
Bagwell, Jeff	7.945	HOU	1	7	535	99.25	0	0
Bonds, Barry	8.917	SF	1	12	551	98.44	1	0
Boone, Bret	2.800	CIN	1	5	398	98.88	0	0
Bordick, Mike	3.583	BAL	0	7	327	97.89	0	0
Damon, Johnny	0.400	KC	0	2	387	98.63	0	0
Galarraga, Andres	8.000	ATL	1	12	494	99.17	0	1
Griffey, Ken Jr.	7.979	SEA	0	9	560	98.62	1	0
Grissom, Marquis	5.000	MIL	1	8	413	98.75	1	0
·	·	·	·	·	·	·	·	·
·	·	·	·	·	·	·	·	·
·	·	·	·	·	·	·	·	·

as type: to Microsoft Excel Workbook. Saving the file as a Microsoft Excel Workbook file will allow you to use the file at a later date without having to go through the steps given previously.

Transforming Variables

In many cases, we may wish to transform an existing variable into a new one. For example, in Chapter 5, we considered adding YEARS SQUARED to our list of independent variables in our MLB player salary model [see (5.2)]. This variable is created by calculating the square of the variable YEARS. This can be performed in Excel quite easily by following these steps:

1. Create a new column heading for the new variable (e.g., YEARS SQUARED).
2. In the cell immediately below this new heading, we may enter a formula. The formula should be typed literally as the following: $= (E2 * E2)$. Then hit Enter. This will take the value shown in cell E2 (which is the first entry of data for the column headed YEARS). Multiply it by itself and put the resulting value in the current cell. This formula can then be copied from this cell to the cells below (for the entire length of the data set) using the Copy and Paste commands. This will create a column of squared values for the variable YEARS, which can be used in our regressions.

Other transformations can be made in a similar way, for example, calculating the natural log for a column of data. See Excel's help index for a list of formulas that can be used.

Performing a Least-Squares Regression

Having inputted the data and created the variables needed for our sample regression model, we are now ready to carry out a least-squares regression. Our first step is to determine which variable will be our dependent variable and which one(s) will be our independent variable(s). Again, working with our baseball example, we will have SALARY as our dependent variable. For independent variables, we can use YEARS, SLUGGING, FIELDING, and YEARS SQUARED (we will omit the variables for NL, BLACK, and HISPANIC for this example). Before performing the regression, one restriction of Excel's regression program is that all columns containing data for the independent or X variables must be alongside each other. That is, we cannot have the columns arranged as shown in Table B.2(A) because the variables BLACK and HISPANIC are not included in the regression. The columns with BLACK and HISPANIC must be moved so that the columns for YEARS, SLUGGING, FIELDING, and YEARS SQUARED can be alongside each other as shown in Table B.2(B). (Note that, in moving columns around, care must be taken so that formulas that are in cells continue to represent the proper equation. As columns are moved from one location to another, the cell formulas adjust cell addresses to the new relative location and thus the resulting equation may not be the one intended.)

Now with our columns properly aligned, we are ready to perform a regression analysis. To do so, we follow these steps:

1. From the main menu bar, select Tools, then Data Analysis ..., and then Regression.[4]
2. You will now be looking at the Regression screen. At this screen, you need to tell the program where the data for the dependent and independent variables are located. Click on the box for Input Y Range: so that there is a blinking cursor there. Next, move the arrow to the column headed SALARY and, starting with the cell containing the variable name SALARY, click and drag down this column, thus highlighting it. As you do so, you should see the cell addresses for the location of the dependent variable appearing in the Input Y Range: box. The location of the dependent variable data has now been completed.
3. Return to the Regression screen and click on the box for Input X Range, thus putting a blinking cursor there. Then take the arrow and now, starting with the cell containing the variable name YEARS, click and drag over to YEARS SQUARED and then down to end of that column. Doing so should highlight the block that contains all the data for the independent variables, including the column headings. As this is done, the address of this block should appear in the Input X Range: box. The location of the independent variable data has now been completed.
4. Next, check the box headed Labels. This tells the program that the first row of data contains the variable names.

TABLE B.2

(A)

PLAYER	SALARY	TEAM	NL	YEARS	SLUGGING	FIELDING	BLACK	HISPANIC	YEARS SQUARED
Abreu, Bob	0.18	PHI	1	1	372	97.92	1	0	1
Anderson, Brady	5.442	BAL	0	11	434	98.78	0	0	121
Bagwell, Jeff	7.945	HOU	1	7	535	99.25	0	0	49
Bonds, Barry	8.917	SF	1	12	551	98.44	1	0	144
Boone, Bret	2.800	CIN	1	5	398	98.88	0	0	25
Bordick, Mike	3.583	BAL	0	7	327	97.89	0	0	49
Damon, Johnny	0.400	KC	0	2	387	98.63	0	0	4
Galarraga, Andres	8.000	ATL	1	12	494	99.17	0	1	144
Griffey, Ken Jr.	7.979	SEA	0	9	560	98.62	1	0	81
Grissom, Marquis	5.000	MIL	1	8	413	98.75	1	0	64
·	·	·	·	·	·	·	·	·	·
·	·	·	·	·	·	·	·	·	·
·	·	·	·	·	·	·	·	·	·

TABLE B.2 (Continued)

(B)

PLAYER	SALARY	TEAM	NL	YEARS	SLUGGING	FIELDING	YEARS SQUARED	BLACK	HISPANIC
Abreu, Bob	0.18	PHI	1	1	372	97.92	1	1	0
Anderson, Brady	5.442	BAL	0	11	434	98.78	121	0	0
Bagwell, Jeff	7.945	HOU	1	7	535	99.25	49	0	0
Bonds, Barry	8.917	SF	1	12	551	98.44	144	1	0
Boone, Bret	2.800	CIN	1	5	398	98.88	25	0	0
Bordick, Mike	3.583	BAL	0	7	327	97.89	49	0	0
Damon, Johnny	0.400	KC	0	2	387	98.63	4	0	0
Galarraga, Andres	8.000	ATL	1	12	494	99.17	144	0	1
Griffey, Ken Jr.	7.979	SEA	0	9	560	98.62	81	1	0
Grissom, Marquis	5.000	MIL	1	8	413	98.75	64	1	0
⋮	⋮	⋮	⋮	⋮	⋮	⋮	⋮	⋮	⋮

5. Under the Residuals option, check the box for Residuals. This will save our residuals from the regression for later use.

6. Select OK. The regression results will be produced on a separate sheet.[5]

Using SPSS[6]

The program SPSS is better suited for regression analysis than Excel. Once data are read into SPSS, it is easier to transform the data and to perform regression analysis. In addition, SPSS is capable of performing more sophisticated regression techniques.

Inputting Data

By Hand

Initiating the program brings you to the SPSS Data Editor. As with Excel, data should be entered as columns, where each column contains all observations on a single variable. Data can be simply typed in and a column can contain nonnumerical entries, such as player names. When a column is started, SPSS assigns the column a generic name (e.g., var000 . . .). To change the column name into something more understandable (e.g., SALARY), click on the tab at the bottom of the Data Editor screen called Variable View. At this screen, you can rename the column giving it a Name (of no more than eight characters) and also a Label (which can have more than eight characters). This can be done for each column of data.

Reading Data from a Data File

To read data from a data file, it is assumed that the text data file has been downloaded and saved on a disk.[7] As noted earlier, these instructions are for the data files that are available on line, which have been saved as tab delimited text files.[8] Other data files may require different settings when reading them into SPSS. To read the text data file into SPSS, we follow these steps:

1. On the Data View page of the SPSS Data Editor, select File, then Read Text Data. Choose the appropriate location for Look in. Pro-

vide the appropriate File Name: (e.g., bball.txt), and under Files of type: select Text Files, then select Open.

2. The next screen is SPSS's Text Import Wizard. On the first page of this wizard you are asked: Does your text file match a predefined format? Select No (the default) and then select Next.

3. On the next page, you are asked: How are your variables arranged? Select Delimited (the default). The next question is: Are variable names included at the top of your file? Select Yes and then select Next.

4. On the third page of the Text Import Wizard, there are a series of questions, all for which the default settings can be used (i.e., the first case of data begins on line 2, each line represents a case, and we want to import all of the cases). Thus, we can simply select Next.

5. The fourth page of the Wizard asks: Which delimiters appear between variables? The default selections are Tab, Comma, and Space. *Unselect* Comma and Space, leaving only Tab as the selection. Select Next.

6. No changes are needed on page 5 of the Wizard. Thus, select Next.

7. The default settings on page 6 of the Wizard can be used (i.e., No and No). Thus, we may select Finish. The data should now be shown on the Data View page of the SPSS Data Editor. Note that if we select Variable View we may add Labels for each variable by simply typing them in. This may be useful because variable Names are restricted to only eight characters, whereas Labels may have more than eight characters. When regressions are performed, Labels (if they are given) are used in the results produced, which may make them easier to read.

Transforming Variables

Suppose we wish to calculate the natural log of SALARY for use in a regression analysis (see Problem 5.1). To create a new variable that is a function of existing variable(s), we may use the following steps:

1. While on the SPSS Data Editor page, choose Transform, then select Compute ....

2. You will next see a page called Compute Variable. At the top, you see a box titled Target Variable. In this box, you must type the

name of the new variable you wish to create, for example, lnsalary
for the natural log of salary (note that the name of this variable is
restricted to eight characters; you may add a variable Label later
as noted previously in step 7).

3. In the box titled Numeric Expression, you must give the equation
that will be used to calculate the Target Variable. In our example,
we would type the expression: ln(salary). The ln is the command
for calculating the natural log of the variable that appears in the
parentheses. (Note that rather than typing salary, we may instead
highlight the variable shown in the list of variables on the left-
hand side of the box, and then click on the Arrow button.)

4. Finally, we select OK. The newly created variable is now shown as
a new column of data on the SPSS Data Editor page.

Performing a Least-Squares Regression

Given that our data have been successfully read into SPSS and we
have created the variables we need for our regression, we now follow
these steps to perform a least-squares regression:

1. At the SPSS Data Editor page, choose Analyze, then Regression,
then select Linear .... This brings us to the Linear Regression page.
At this page, there are many selections we can make. We will use
only the ones necessary to perform the simple least-squares regression used in this book, including saving the residuals from
the regression.

2. From the list of variables shown on the left-hand side, we must
select our dependent variable. We may choose, for this example,
salary. To select salary as our dependent variable, we may simply
click on this variable shown in the variable list, thus highlighting
it, and then click on the Arrow button. The variable will appear in
the Dependent: box. (Note that, to *unselect* a variable, we may click
on the Arrow button again. Further, we may type variable names
in boxes, rather than using the Arrow button.)

3. Our next task is to choose our independent variable(s). To do so,
we may simply highlight the variables shown in our variable list
and then click on the Arrow button. Doing so, the variables are
then placed in the box titled Independent(s).

4. To save the residuals from our regression, we select Save ... shown at the bottom of the Linear Regression page. This brings us to the Linear Regression: Save page. Select the box for Unstandardized under Residuals and then select Continue.

5. Select OK. The regression is performed and the output is produced on an Output1 page in the SPSS Viewer.[9] (Note that you may toggle back and forth from the SPSS Viewer and the SPSS Data Editor by clicking on the appropriate buttons at the bottom of the page or by minimizing screens.)

————•◆•————

▼ **Notes**

1. There are a number of books that describe how to use Excel and SPSS. See, for example, Berk and Carey (2000) for Excel and Einspruch (1998) for SPSS.

2. Data are available at www.sbeusers@csuhayward.edu/-lkahane/index. html.

3. In fact, data input errors are another source of stochastic error, e_i, in our sample regression model.

4. Note that if you do not see Data Analysis under the Tools menu, this is likely because this particular module was not loaded during the installation of the Excel program. To load this module now, go to Tools, select Add-Ins ... , and check the boxes for Analysis Toolpak and Analysis Toolpak - VBA. Doing this will load the module containing the Data Analysis features.

5. Note that the regression results presented will differ somewhat to the ones shown in the text of this book. The results shown in the text were edited to simplify our discussion.

6. These instructions are for SPSS Version 10.0. Other versions may have slightly different steps required to perform the tasks described in this appendix.

7. It should be noted that SPSS has the ability to read various file formats, including files that have been saved as Excel files. For instructions on how to read files of different formats, see the help index in SPSS.

8. See note 2.

9. The output produced will be slightly different from that shown in this book. SPSS output provided in this book was edited to simplify our discussion.

APPENDIX C
t TABLE

TABLE C.1

df	Confidence Level Probability	50% 0.5	80% 0.2	90% 0.1	95% 0.05	98% 0.02	99% 0.01
1		1	3.078	6.314	12.706	31.821	63.657
2		0.816	1.886	2.92	4.303	6.965	9.925
3		0.765	1.638	2.353	3.182	4.541	5.841
4		0.741	1.533	2.132	2.776	3.747	4.604
5		0.727	1.476	2.015	2.571	3.365	4.032
6		0.718	1.44	1.943	2.447	3.143	3.707
7		0.711	1.415	1.895	2.365	2.998	3.499
8		0.706	1.397	1.86	2.306	2.896	3.355
9		0.703	1.383	1.833	2.262	2.821	3.25
10		0.7	1.372	1.812	2.228	2.764	3.169
11		0.697	1.363	1.796	2.201	2.718	3.106
12		0.695	1.356	1.782	2.179	2.681	3.055
13		0.694	1.35	1.771	2.16	2.65	3.012
14		0.692	1.345	1.761	2.145	2.624	2.977
15		0.691	1.341	1.753	2.131	2.602	2.947
16		0.69	1.337	1.746	2.12	2.583	2.921
17		0.689	1.333	1.74	2.11	2.567	2.898
18		0.688	1.33	1.734	2.101	2.552	2.878
19		0.688	1.328	1.729	2.093	2.539	2.861
20		0.687	1.325	1.725	2.086	2.528	2.845
21		0.686	1.323	1.721	2.08	2.518	2.831
22		0.686	1.321	1.717	2.074	2.508	2.819
23		0.685	1.319	1.714	2.069	2.5	2.807
24		0.685	1.318	1.711	2.064	2.492	2.797
25		0.684	1.316	1.708	2.06	2.485	2.787
26		0.684	1.315	1.706	2.056	2.479	2.779
27		0.684	1.314	1.703	2.052	2.473	2.771
28		0.683	1.313	1.701	2.048	2.467	2.763
29		0.683	1.311	1.699	2.045	2.462	2.756
30		0.683	1.31	1.697	2.042	2.457	2.75
40		0.681	1.303	1.684	2.021	2.423	2.704
50		0.679	1.299	1.676	2.009	2.403	2.678
60		0.679	1.296	1.671	2	2.39	2.66
70		0.678	1.294	1.667	1.994	2.381	2.648
80		0.678	1.292	1.664	1.99	2.374	2.639
90		0.677	1.291	1.662	1.987	2.368	2.632
100		0.677	1.29	1.66	1.984	2.364	2.626
∞		0.674	1.282	1.645	1.96	2.326	2.576

NOTE: The values in the table are the t values for the given areas in both tails.

APPENDIX D
ANSWERS TO PROBLEMS

Chapter 1

1.1 a. The intercept, α, is the expected value of the dependent variable (Y) for the case when the independent variable (X) has a value of 0. In this example, the intercept is the expected wage for an individual with zero years of education. The expected sign for α, in this case, is a positive number, indicating that someone with no education, but working will earn a positive wage.

b. The coefficient to X, β, represents the average effect on Y for a one-unit increase in X. In this case, it shows how an individual's wage is expected to change for each additional year of education. We expect the sign for β to be positive, indicating that, as an individual's education level increases, so should his or her wage, all else being equal.

c. Recall that the error term, u_i, captures all other effects that are not taken into account by our model. Thus, in this case, things that might affect an individual's wage such as work experience, gender, race, and so on may be present in the error term.

1.2 There are many factors that may affect the outcome of a presidential election besides the real growth rate. For example, other economic measures (e.g., inflation or unemployment) might have an impact on the number of votes that the incumbent-party candidate receives. In addition, there could be advantages for presidents who run for reelection. These issues are taken up later Chapters 4 and 5.

Chapter 2

2.1 To answer this question, we can first calculate $\overline{X}$ and $\overline{Y}$, which are simply the average values of each variable. Adding the X column and dividing by the sample size of 5, we obtain: $61/5 = 12.2$. Doing the same for the Y column, we obtain: $50.5/5 = 10.1$. Using these values, we can now construct the following table:

X_i	Y_i	$(X_i - \overline{X}) * (Y_i - \overline{Y})$	$(X_i - \overline{X})^2$
13	10.5	0.32	0.64
12	9.75	0.07	0.04
12	10.0	0.02	0.04
14	12.25	3.87	3.24
10	8.00	4.62	4.84
Sum: 61	50.5	8.9	8.8

Using the preceding table and (2.3a), we have for b:

$$b = \frac{\sum_{i=1}^{n}(X_i - \overline{X})(Y_i - \overline{Y})}{\sum_{i=1}^{n}(X_i - \overline{X})^2} = \frac{8.9}{8.8} = 1.011.$$

Using (2.3b), we have for the intercept a:

$$a = \overline{Y} - b\overline{X} = 10.1 - (1.011 * 12.2) = -2.23.$$

Given these values for a and b, we can write the predicted equation as

$$\widehat{Y}_i = -2.23 + 1.011X_i.$$

Thus, according to the predicted equation, if the value of X_i is 0, then the predicted value for the wage, Y_i, is approximately $-\$2.23$ per hour. The interpretation of the intercept, in this case, is obviously meaningless because a negative wage is not possible. As for b, the coefficient to X_i, the value 1.011 tells us that each additional year of education tends to increase an individual's hourly wage by approximately $1.01 per hour, other things being equal.

2.2 Table A.1 in Appendix A reports that Mark McGwire had 10 years of MLB experience by the end of 1997. Plugging this value into (2.5b) produces the following predicted salary:

$$\widehat{SALARY}_i = 2.031 + 0.350(10)_i = 5.531.$$

Comparing this prediction to McGwire's actual salary of $8.333 million, we find that the error in prediction is

$$e_i = Y_i - \widehat{Y}_i = 8.333 - 5.531 = 2.802.$$

Thus, the OLS regression equation underpredicts McGwire's salary by approximately $2.8 million. This sizable error is likely due to the fact that our two-variable regression model is overly simplistic as it does not take into account other important factors (such as his offensive and defensive abilities) that may explain Mark McGwire's salary. Chapter 4 introduces the multiple regression model, which allows us to build more realistic models that include more than one X variable as a predictor for the dependent variable, Y.

2.3 a. The literal interpretation of the intercept, α, is the expected price of an SUV with zero horsepower. Obviously, however, this interpretation has no practical meaning because a vehicle with zero horsepower is not typically sold. As such, we have no meaningful expectation for the sign of α. The slope term, β, would show us the

TABLE D.1

		Coefficients[a]			
		Unstandardized Coefficients			
Model		B	Std. Error	t	Sig.
1	(Constant)	−2762.283	59333.133	−0.466	0.644
	HORSEPOWER	174.958	29.698	5.891	0.000

[a]Dependent variable: MANUFACTURER'S SUGGESTED RETAIL PRICE.

expected increase in the price of an SUV for a one-unit increase in horsepower. Given that consumers would prefer more powerful cars, other things being equal, we would expect a positive value for β.

b. The SPSS results are shown in Table D.1. As noted, the intercept term has no reasonable interpretation. The coefficient for HORSEPOWER is 174.958, which tells us that, on average, an SUV's price increases by about $175 for a one-unit increase in horsepower.

Chapter 3

3.1 The Excel output is shown in Table D.2. The intercept term is approximately −5.13, meaning that a person with no education is expected to earn −$5.13 per hour. Obviously, this interpretation is not sensible. It should be noted that, in the data set shown in Table A.6, none of the observations has a zero value for education. Thus, even though our intercept term is technically required to define our sample regression line, it has no practical meaning. The estimated coefficient to education (EDUC) is 1.08, indicating that, as education increases by 1 year, an individual's salary is expected to increase by about $1.08, other things being equal. The R^2 is reported as approximately 0.261, meaning that about 26.1% of the behavior (i.e., the variation) of Y is explained by our model; a rather poor fit.

3.2 a. Referring to Table D.2, the estimated intercept term has a P value of approximately 0.163. Thus, because the P value is greater than the chosen level of significance for this test (i.e., $0.163 > 0.05$), this means that we cannot reject the hypothesis, H_0: $\alpha = 0$. In other words, we cannot reject the hypothesis that the population's intercept term is equal to 0 at the 5% significance (95% confidence) level.

b. Table D.2 reports the P value for the coefficient to education as approximately 0.0003. Because this value is less than the chosen level of significance (i.e., $0.0003 < 0.01$), we can reject the hypothesis, H_0: $\beta = 0$. That is, we can reject the hypothesis that the population's slope term is equal to 0 at the 1% significance (99% confidence) level.

TABLE D.2

Summary Output

Regression Statistics

Multiple R	0.51094344
R Square	0.261063199
Adjusted R Square	0.243878623
Standard Error	4.923320483
Observations	45

ANOVA

	df	SS	MS	F	Significance F
Regression	1	368.2332743	368.2332743	15.19171539	0.000335184
Residual	43	1042.280637	24.23908458		
Total	44	1410.513911			

	Coefficients	Standard Error	t Statistic	P Value
Intercept	−5.127272985	3.611264708	−1.419799821	0.162875679
EDUC	1.084226022	0.278173925	3.897655115	0.000335184

3.3 Table 3.2 reports the coefficient for the intercept term as approximately 51.52 with an estimated standard error of about 1.099. Using these values and (3.6b), where h in this case is 50, we have

$$t \text{ statistic} = \frac{b-h}{se(b)} = \frac{51.52-50}{1.099} = 1.383.$$

The output shown in Table 3.2 reports that 21 observations are used in the regression. This means that there are 19 degrees of freedom left over (equal to the number of observations minus the number of estimated coefficients) for our hypothesis test. Viewing Table C.1, we go down the column headed df (short for degrees of freedom) to row 19. We move across to the column headed 0.05 (given our significance level for the test is 5%) and find the t value 2.093. This is the value to which we compare the preceding t statistic. Because |t statistic| < t value (i.e., 1.383 < 2.093), this means that we cannot reject H_0: $\alpha = 50$ at the 5% significance (95% confidence) level. That is, we cannot reject the hypothesis that the population's true intercept is equal to 50.

Chapter 4

4.1 Table D.3 shows the SPSS output of the regression shown in (4.11), including the variable EDUC (the percentage of a state's population who is 25 years or older with a high school degree or equivalent) as an independent variable.

a. The estimated coefficient for EDUC is −0.310, meaning that, from one state to another, as the percentage of a state's population, who is 25 years or older with a high school degree (or equivalent) increases by 1%, the abortion rate tends to decrease by approximately −0.310, other things being equal.

b. Because the regression estimated in Table D.3 has a different number of independent variables as compared that in (4.11), the adjusted R^2 is the measure that we can use to compare the overall performance of the models. The adjusted R^2 for the model in (4.11) is shown in Table 4.3 and is reported as 0.498. Table D.3 reports an adjusted R^2 of 0.514. Thus, based on the adjusted R^2 measures, the model that includes EDUC is superior.

TABLE D.3

		Model Summary		
Model	R	R Square	Adjusted R Square	Std. Error of the Estimate
1	0.751[a]	0.563	0.514	7.0136196

[a]Predictors: (Constant), EDUC, RELIGION, PRICE, PICKETING, INCOME.

		ANOVA[b]				
Model		Sum of Squares	df	Mean Square	F	Sig.
1	Regression	2793.228	5	558.646	11.357	0.000[c]
	Residual	2164.398	44	49.191		
	Total	4957.626	49			

[b]Dependent variable: ABORTION RATE.

[c]Predictors: (Constant), EDUC, RELIGION, PRICE, PICKETING, INCOME.

		Coefficients[d]			
		Unstandardized Coefficients			
Model		B	Std. Error	t	Sig.
1	(Constant)	12.584	14.824	0.849	0.401
	RELIGION	$-1.60E-03$	0.082	-0.020	0.984
	PRICE	$-4.05E-02$	0.022	-1.846	0.072
	INCOME	$2.627E-03$	0.000	6.430	0.000
	PICKETING	-12.560	4.077	-3.081	0.004
	EDUC	-0.310	0.197	-1.572	0.123

[d]Dependent variable: ABORTION RATE.

c. Reviewing the output shown in Table D.3, the "Sig." value (analogous to Excel's *P* value) for the coefficient to EDUC is 0.123. Given a significance level of 5%, because 0.123 > 0.05, we cannot reject the hypothesis that the population's coefficient to EDUC is truly 0. That is, we cannot reject the hypothesis that, for the population data, EDUC has no effect on the abortion rate.

4.2 Table 4.2 shows the estimated coefficient for GROWTH to be 0.700 with a standard error of 0.203. Thus, using (3.6b) to calculate the appropriate t statistic, we find

$$t \text{ statistic} = \frac{b-h}{se(b)} = \frac{0.700 - 1.0}{0.203} = -1.478,$$

where h, the hypothesized value for the coefficient to GROWTH, is 1.0. Given a significance level of 10% and 18 degrees of freedom (21 observations minus 3 estimated parameters), the t value for the t tables is shown as 1.734. Thus, because $|-1.478| < 1.734$, we cannot reject the hypothesis that the coefficient to GROWTH for the population's data is equal to 1.0.

4.3 Plugging 10 for YEARS, 380 for SLUGGING, and 97.5 for FIELDING into (4.7b), we find the predicted salary to be

$$\widehat{SALARY}_i = -76.234 + 0.291(10) + 0.022(380) + 0.695(97.5) = 2.798.$$

That is, a player with 10 years of MLB experience, a career slugging average of 380, and a career fielding percentage of 97.5 is expected to make about $2.8 million, according to our sample regression results.

4.4 a. The SPSS output for the regression is shown in Table D.4. Using the estimated coefficients provided, we can write the predicted equation, after rounding to two decimal places, as

$$\widehat{SALARY}_i = -9.40 + 1.26(EDUC_i) + 0.11(EXPER_i).$$

Interpreting our results, we see immediately that the intercept term, estimated as −9.40, is not sensible and warrants no interpretation. Again, the intercept is technically required to anchor our line, but given that EDUC and EXPER do not take on zero values in our data set, the intercept value is not meaningful. As for the coefficient to EDUC, the value of 1.26 suggests that, on average, an additional year of education raises the hourly wage by $1.26, other things being equal. For EXPER, the coefficient of 0.11 implies that, on average, an additional year of experience raises hourly wages by about 11 cents, all else being equal.

TABLE D.4

		Model Summary		
Model	R	R Square	Adjusted R Square	Std. Error of the Estimate
1	0.553[a]	0.306	0.273	4.8276

[a]Predictors: (Constant), EXPER, EDUC.

			ANOVA[b]			
Model		Sum of Squares	df	Mean Square	F	Sig.
1	Regression	431.688	2	215.844	9.262	0.000[c]
	Residual	978.826	42	23.305		
	Total	1410.514	44			

[b]Dependent variable: WAGE.

[c]Predictors: (Constant), EXPER, EDUC.

		Coefficients[d]			
		Unstandardized Coefficients			
Model		B	Std. Error	t	Sig.
1	(Constant)	−9.399	4.386	−2.143	0.038
	EDUC	1.263	0.293	4.303	0.000
	EXPER	0.110	0.067	1.650	0.106

[d]Dependent variable: WAGE.

b. Observing the "Sig." values shown in Table D.4, we have 0.000 for EDUC. Thus, we can reject the hypothesis H_0: $\beta_1 = 0$ at the 5% significance level (i.e., 0.000 < 0.05). As for EXPER, we have a "Sig." value of 0.106. Thus, we cannot reject the hypothesis H_0: $\beta_2 = 0$ at the 5% significance level (i.e., 0.106 > 0.05).

4.5 a. The Excel regression results are shown in Table D.5. As we can see, the negative coefficient to the intercept term has no sensible

TABLE D.5

Summary Output

Regression Statistics

Multiple R	0.721008462
R Square	0.519853202
Adjusted R Square	0.492416242
Standard Error	8085.12336
Observations	38

ANOVA

	df	SS	MS	F	Significance F
Regression	2	2477125626	1238562813	18.94718674	2.65488E − 06
Residual	35	2287922691	65369219.74		
Total	37	4765048317			

	Coefficients	Standard Error	t Statistic	P Value
Intercept	−2203.122611	5856.0565	−0.376212663	0.709029761
horsepower	112.1389418	52.17517503	2.14927773	0.038606975
engine size (liters)	3192.849302	2195.988244	1.453946446	0.154870326

interpretation. The coefficient to horsepower, however, is positive as expected and, given its P value of approximately 0.039, it is statistically different from 0 at the 5% level of significance (i.e., $0.039 < 0.05$). The estimated coefficient is approximately 112.14, implying that, on average, one additional unit of horsepower increases the MSRP by about $112, all else being equal.

Regarding the engine size, the P value is about 0.155, indicating that this coefficient is *not* statistically different from 0 at the 5% significance level (i.e., $0.155 > 0.05$). As a matter of practice, however, we can interpret its estimated coefficient. The coefficient is approximately 3192.85, indicating that, on average, the MSRP increases by over $3000 for each liter increase in engine size, all else being equal.

b. Given that we have a different number of independent variables in the two models, the adjusted R^2 is the appropriate comparison (not the R^2). In the model with engine size, the adjusted R^2 is 0.492, which is greater than the model without engine size (with an adjusted R^2 of 0.477), indicating that including engine size produces a better fit.

Chapter 5

5.1 a. The regression results are shown in Table D.6. The predicted equation is

$$\ln\widehat{SALARY}_i = -32.71 + 0.149(YEARS_i)$$
$$+ 6.737E - 3(SLUGGING_i) + 0.301(FIELDING_i).$$

b. The coefficient to YEARS (years of MLB experience) is shown as 0.149. Multiplying this by 100, we have 14.9, which means that an additional year of experience in MLB tends to increase salary by approximately 14.9%, other things being equal. The coefficient to SLUGGING (slugging average) is $6.737E - 3$ (which is in scientific notation). Multiplying this coefficient by 100, we have 0.6737. Thus, an increase in the slugging average of one point leads to approximately a 0.67% increase in salary, all else being equal. Finally, the coefficient to FIELDING (fielding percentage) is 0.301, which means that a 1% increase in fielding percentage increases salary by approximately 30.1%, other things being equal.

TABLE D.6

Model Summary

Model	R	R Square	Adjusted R Square	Std. Error of the Estimate
1	0.809[a]	0.655	0.618	0.7317

[a]Predictors: (Constant), FIELDING, SLUGGING, YEARS.

ANOVA[b]

Model		Sum of Squares	df	Mean Square	F	Sig.
1	Regression	28.490	3	9.497	17.736	0.000[c]
	Residual	14.993	28	0.535		
	Total	43.482	31			

[b]Dependent variable: NATURAL LOG OF SALARY.

[c]Predictors: (Constant), FIELDING, SLUGGING, YEARS.

Coefficients[d]

Model		Unstandardized Coefficients		t	Sig.
		B	Std. Error		
1	(Constant)	−32.710	10.778	−3.035	0.005
	YEARS	0.149	0.029	5.084	0.000
	SLUGGING	$6.737E-03$	0.002	3.306	0.003
	FIELDING	0.301	0.109	2.756	0.010

[d]Dependent variable: NATURAL LOG OF SALARY.

5.2 a. Table D.7 shows the SPSS regression results. The coefficient to FUNDS is 3.023, meaning that states that provide public funds for abortion services tend to have an abortion rate that is 3.023 higher than states that do not provide public funds for these services, all else being equal. Similarly, the coefficient to LAWS is −1.163, suggesting that states with laws that restrict access to abortion services (such as parental consent laws) and that are enforced tend

TABLE D.7

		Model Summary		
Model	R	R Square	Adjusted R Square	Std. Error of the Estimate
1	0.746[a]	0.556	0.494	7.1521707

[a]Predictors: (Constant), FUNDS, RELIGION, PRICE, PICKETING, LAWS, IN-COME.

			ANOVA[b]			
Model		Sum of Squares	df	Mean Square	F	Sig.
1	Regression	2758.023	6	459.671	8.986	0.000[c]
	Residual	2199.602	43	51.154		
	Total	4957.626	49			

[b]Dependent variable: ABORTION RATE.

[c]Predictors: (Constant), FUNDS, RELIGION, PRICE, PICKETING, LAWS, IN-COME.

		Coefficients[d]			
		Unstandardized Coefficients			
Model		B	Std. Error	t	Sig.
1	(Constant)	−2.586	9.622	−0.269	0.789
	RELIGION	$2.622E-02$	0.087	0.300	0.766
	PRICE	$-4.65E-02$	0.022	−2.087	0.043
	INCOME	$2.156E-03$	0.000	5.038	0.000
	PICKETING	−9.986	4.105	−2.433	0.019
	LAWS	−1.163	2.398	−0.485	0.630
	FUNDS	3.023	2.815	1.074	0.289

[d]Dependent variable: ABORTION RATE.

to have an abortion rate that is 1.163 *less* than states that either do not have such laws or do not enforce them, other things being equal.

b. Because the model estimated in this problem has a greater number of X variables as compared to that estimated in (4.11), the *adjusted* R^2 should be used to compare these models. Table 4.3 reports an adjusted R^2 of 0.498, whereas the adjusted R^2 for the regression shown in Table D.7 is 0.494. Thus, by this measure, the model that includes the dummy variables does not outperform the original one. (Note: The R^2 for the present regression is 0.556, which is larger than that shown in Table 4.3, reported as 0.539. Thus, this problem provides a good demonstration of the point raised in Chapter 4 that the R^2 may increase even if we add X variables to our model that are apparently not statistically important. This example illustrates why the adjusted R^2 should be used for comparing models that have a different number of X variables.)

c. Noting the "Sig." values for the estimated coefficients for FUNDS and LAWS (shown as 0.630 and 0.289, respectively), in both cases they are larger than our significance level (i.e., 0.05). Thus, we cannot reject the hypothesis that they are statistically equal to 0. (This is not surprising, given our previous answer to part *b*.)

5.3 The regression results are reported in Table D.8. The coefficient to the time index is approximately −0.164, which indicates that the percentage of two-party votes that the incumbent-party candidate receives tends to decrease by about 0.164% from one election period to another, other things being equal. Comparing this model to that without the time index, we see that the adjusted R^2 for this model is shown as 0.656, whereas the adjusted R^2 for the model without the time index is shown as 0.655 (see Table 5.2). Thus, the model with the time index included is marginally superior to that without the index, judging by the adjusted R^2's. The estimated coefficient to the time index, however, is not statistically different from 0 at the 10% level of significance because the P value (approximately 0.313) is greater than our chosen significance level (0.10). Thus, it appears that the time index is not an important independent variable in this regression.

5.4 The OLS results are shown in Table D.9. We can see from the output that, with the exception of the intercept and the coefficient

TABLE D.8

Summary Output

Regression Statistics

Multiple *R*	0.851592608
R Square	0.72520997
Adjusted *R* Square	0.656512462
Standard Error	4.17361665
Observations	21

ANOVA

	df	SS	MS	F	Significance F
Regression	4	735.5426992	183.8856748	10.55656887	0.000221118
Residual	16	278.705215	17.41907594		
Total	20	1014.247914			

	Coefficients	Standard Error	*t* Statistic	*P* Value
Intercept	52.96417426	2.842968869	18.6298819	2.84965E – 12
Growth	0.729031521	0.18964268	3.962897408	0.00115786
Inflation	−0.547761587	0.289293621	−1.893445092	0.076527533
Incumbent	4.687069866	1.936896719	2.419886316	0.027795848
Time	−0.164166699	0.157682913	−1.04119144	0.313297591

TABLE D.9

Summary Output

Regression Statistics

Multiple R	0.841442759
R Square	0.708025917
Adjusted R Square	0.635032396
Standard Error	4.302136861
Observations	21

ANOVA

	df	SS	MS	F	Significance F
Regression	4	718.1138092	179.5284523	9.699846073	0.000351968
Residual	16	296.1341051	18.50838157		
Total	20	1014.247914			

	Coefficients	Standard Error	t Statistic	P Value
Intercept	51.46156097	2.642020148	19.47811072	1.43797E−12
Growth	0.573147258	0.443273961	1.292986523	0.214374642
Inflation	−0.588368958	0.370668825	−1.587240007	0.13202133
Incumbent	4.608812837	2.04072917	2.258414739	0.03823958
Growth*Incumbent	0.127081849	0.45369 2498	0.280105688	0.782985532

to INCUMBENT, all of the P values are quite large and thus we could not reject the hypothesis that the population coefficients are 0 for the other variables. As a matter of practice, though, we can interpret all the estimated coefficients. Starting with the intercept, we find that if all other variables took on a zero value (i.e., zero growth and inflation, and a nonincumbent), then the incumbent-party candidate would be expected to receive 51.46% of the two-party vote. The effects of economic growth on VOTES has two sources: the direct effect, shown as the coefficient to GROWTH, 0.573; and the coefficient for the interaction effect between GROWTH and INCUMBENT, 0.127. Thus, if the economy grows by 1%, then the incumbent-party candidate is expected to gain about 0.573% of the two-party vote. However, if the incumbent-party candidate is also the incumbent, then the effect of a 1% increase in growth is equal to the sum of the GROWTH coefficient and the coefficient for the interaction of GROWTH and INCUMBENT. Thus, we have $0.573 + 0.127 = 0.7$, or a 0.7% increase in the percentage of two-party votes. We have a similar interpretation of the effects of being the actual incumbent. The coefficient to INCUMBENT is shown as 4.609, implying that all else being equal, incumbents enjoy about a 4.6% increase in the percentage of two-party votes, as compared to nonincumbents. However, the *full effect* of being the incumbent also includes the interaction effect. Thus, we have $4.609 + 0.127(\text{GROWTH})$. In this case, if the economy grows by 1% and the incumbent-party candidate is, in fact, the incumbent, then the expected gain in the percentage of two-party votes is $4.609 + 0.127(1) = 4.736$. Finally, the interpretation of the coefficient to INFLATION, shown as -0.588, is that, all else being equal, the incumbent-party candidate is expected to receive about 0.6% less of the two-party votes for every 1% increase in inflation.

5.5 Using the results shown in Table 5.6 and following the example shown in the text for SLUGGING, a one-unit increase in FIELDING is expected to have the following effect on a player's salary:

$$-7.34041 + 0.01544(\text{SLUGGING}_i).$$

Inserting the mean value for SLUGGING, equal to 474.844, into this expression, we have

$$-7.34041 + 0.01544(474.844) = -0.00882.$$

This result suggests that a one-unit increase in FIELDING actually reduces salary by nearly $9000! As noted in the discussion in Chapter 5, however, we must remember that this result is computed using the average value for SLUGGING. If we consider a player who is an above-average hitter, say with a SLUGGING average of 500, then we have

$$-7.34041 + 0.01544(500) = 0.37959.$$

Or, in this case, a player is expected to increase his salary by about $380,000 for a one-unit increase in FIELDING. As noted in Chapter 5, we see that players who have good offensive *and* defensive skills earn more than players who are skilled in just offense *or* defense.

5.6 a. Given the definition of the dummy variables, the base case (when all dummies have a zero value) is unmarried White (or Asian) males.

b. The Excel regression results are shown in Table D.10. As we have seen before, the negative estimated coefficient for the intercept term has no meaningful interpretation. As for the estimated coefficient for EDUC, we have a value of about 1.14 with an associated P value of approximately 0.0003. These results show that the coefficient is statistically different from 0 at the 10% level of significance (i.e., $0.0003 < 0.10$), and the coefficient implies that, for each added year of education, we expect hourly wage to increase by about $1.14, all else being equal. EXPER has an estimated coefficient of about 0.10, implying that an additional year's worth of experience adds about 10 cents to hourly wage, all else being equal. This interpretation is made with caution, however, as the P value indicates that the estimated coefficient is not statistically different from 0 at the 10% level of significance (i.e., $0.154 > 0.10$). The coefficient to FEMALE is approximately -2.58, indicating that females with the same education and experience as males tend to earn about $2.58 less than males, all else being equal. This estimated coefficient is statistically different from 0 at the 10% level, judging by the P value (i.e., $0.078 < 0.10$). The coefficient to MARRIED

TABLE D.10

Summary Output

Regression Statistics

Multiple R	0.636735011
R Square	0.405431474
Adjusted R Square	0.329204739
Standard Error	4.637216167
Observations	45

ANOVA

	df	SS	MS	F	Significance F
Regression	5	571.8667336	114.3733467	5.318756971	0.000805904
Residual	39	838.6471775	21.50377378		
Total	44	1410.513911			

	Coefficients	Standard Error	t Statistic	P Value
Intercept	-5.999311798	4.483590874	-1.338059597	0.188626902
EDUC	1.138065278	0.28722855	3.962228957	0.000306546
EXPER	0.096605257	0.066502177	1.45266307	0.154316266
FEMALE	-2.578663089	1.422459827	-1.812819625	0.07756224
MARRIED	0.655192718	1.45334511	0.450817025	0.654617997
BLKHISP	-2.941811412	1.798157915	-1.636013938	0.109882898

is 0.65, indicating that married individuals tend to earn about 65 cents more than nonmarried individuals, all else being equal. This coefficient, however, is not statistically significant at the 10% level (i.e., 0.65 > 0.10). Finally, the coefficient to BLKHISP is -2.94, implying that individuals who are Black and/or Hispanic earn about $3 less than White (or Asian) individuals with the same education and experience. This result, however, is somewhat questionable, given that the P value of about 0.11 is greater than 0.10, indicating that the coefficient is not statistically different from 0 at the 10% level of significance.

Chapter 6

6.1 Recall that high-multicollinearity problems occur when two or more independent variables are almost perfectly linearly related. In this case, the inclusion of the independent variable GAMES, equal to the number of career games played, may be closely related to the independent variable YEARS (and to perhaps YEARS SQUARED). The reason for the close relationship is fairly straightforward: The more years a player plays in the major leagues, the more career games he will have played. It is not a perfect linear relationship as the number of games played each year is not likely to be exactly the same from one year to another for all players, but it may be close.

6.2 Table D.11 reports the R^2 values for five auxiliary regressions. As we can clearly see, auxiliary regressions (1), (3), (4), and (5) have relatively high R^2 values, indicating a good fit in each case. Thus,

TABLE D.11

Auxiliary Equation	R^2
(1) HORSEPOWER $= f$(CARGO, MPG, SIZE, CYLINDERS)	0.761
(2) CARGO $= f$(MPG, SIZE, CYLINDERS, HORSEPOWER)	0.543
(3) MPG $= f$(CARGO, SIZE, CYLINDERS, HORSEPOWER)	0.795
(4) SIZE $= f$(CARGO, MPG, CYLINDERS, HORSEPOWER)	0.851
(5) CYLINDERS $= f$(CARGO, MPG, SIZE, HORSEPOWER)	0.892

we find clear evidence of high multicollinearity among these variables.

6.3 Table D.12 shows the SPSS output for the estimated model in (6.5). The positive value for the estimated coefficient to INCOME, reported (in scientific notation) as $4.941E - 04$, agrees with our

TABLE D.12

Model Summary[a]

Model	R	R Square	Adjusted R Square	Std. Error of the Estimate
1	0.333[b]	0.111	0.093	3.9667

[a]Dependent variable: ABSOLUTE VALUE OF RESIDUAL.

[b]Predictors: (Constant), INCOME.

ANOVA[c]

Model		Sum of Squares	df	Mean Square	F	Sig.
1	Regression	94.442	1	94.442	6.002	0.018[d]
	Residual	755.272	48	15.735		
	Total	849.714	49			

[c]Dependent variable: ABBSOLUTE VALUE OF RESIDUAL.

[c]Predictors: (Constant), INCOME.

Coefficients[e]

Model		Unstandardized Coefficients		t	Sig.
		B	Std. Error		
1	(Constant)	−4.135	3.916	−1.056	0.296
	INCOME	$4.941E - 04$	0.000	2.450	0.018

[e]Dependent variable: ABBSOLUTE VALUE OF RESIDUAL.

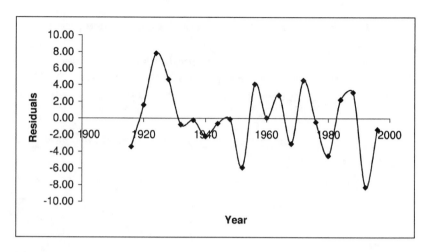

Figure D.1.

graphical analysis of the residuals, namely, that the residuals tend to increase in absolute size as INCOME increases. The t statistic is reported as 2.45 with an associated "Sig." value of 0.018, indicating that the coefficient to INCOME is statistically different from 0 at better than the 5% significance (95% confidence) level. In other words, there appears to be a positive significant relationship between the absolute value of the residuals and INCOME, which is evidence of nonconstant error variance.

6.4 The plot of the residuals from the OLS estimation of (5.4a) is shown in Figure D.1. Although the plot exhibits somewhat of a wave pattern, we do not witness long strings of positive and negative errors and thus there is no obvious autocorrelation problem in this model.

Adjusted R^2	The R^2 adjusted downward to take into account the number of independent variables included in a regression model. It can be used for comparing the goodness of fit of two regression models when they have the same dependent variable, but a different number of independent variables. *See* **R^2**.
Alternative Hypothesis	Used in hypothesis testing, this is the hypothesis that is offered as an alternative case to the "null" hypothesis, the latter of which is typically the focus of a hypothesis test. *See* **Null hypothesis**.
ANOVA	An acronym for *analysis of variance*, which is the breakdown of the total variation of the dependent variable into its two components: the variation explained by the regression and the variation that is unexplained (or the "residual" variation).
Autocorrelation	Also known as "serial correlation," this is the case where the error term from one period in a regression is correlated with the error term from the previous period. If the relationship is a positive one, then this is the case of "positive autocorrelation;" if the errors are negatively correlated, then this is referred to as "negative autocorrelation."

Auxiliary Regression

Used to test for the presence of near-perfect multicollinearity. An independent variable, which is suspected of being a nearly perfect function of one or more other independent variables, is regressed against these independent variables. If the resulting regression is statistically significant (based on an F test), then this provides evidence of near-perfect multicollinearity.

Bivariate Linear Regression Model

See **Two-variable linear regression model**.

BLUE

An acronym for *b*est *l*inear *u*nbiased *e*stimator. An estimator that is BLUE is the "best" in the sense that, compared to all other linear unbiased estimators, the one in question has the smallest variance for the estimated parameter(s).

Cochrane– Orcutt Procedure

A method of estimating a regression model in the presence of autocorrelation. The procedure uses the residuals from the OLS regression estimation to discover the systematic component in the errors and then this information is used to refine the estimates of the coefficients.

Correlation Coefficient

A measure of the degree to which two variables are linearly associated. The coefficient ranges from -1 to $+1$, where a value of -1 means that two variables are perfectly negatively correlated; a value of $+1$ means that they are perfectly positively correlated. A value of 0 means that the two variables are not linearly associated. *See* **Covariance**.

Covariance	A statistical measure showing whether two variables tend to move together. A positive value for covariance indicates that if one variable increases, the other tends to increase, too. A negative value indicates that as one variable increases, the other tends to decrease. (Note: This measure is related to the correlation coefficient. The correlation coefficient, however, shows the *strength* of a linear association, whereas covariance simply shows the *direction* of the relationship.)
Cross-Section Data Set	A set of data that is collected for one point in time. For example, stock prices on a selection of 100 stocks, for a given point in time. *See*: **Time series data set; Pooled data set.**
Degrees of Freedom	Equal to the number of observations in a sample minus the number of estimated parameters. The degrees of freedom represent the remaining amount of information in a sample of data that can be used for other purposes, such as hypothesis testing.
Dependent Variable	In a regression model, it is the variable that we are trying to explain. It is assumed that the dependent variable, sometimes referred to as the Y variable, is a function of the independent variable(s), which are often called the X variable(s).
Dummy Variable	Also called an "indicator" or "categorical" variable. These are variables that are created to indicate whether or not something is true. For example, if a sample of data contains observations on both male and female respondents, a dummy variable can be introduced taking the value of 1 if a respondent is female, 0 in the case of a male

respondent. Note that if there are *m* separate categories, then only $m - 1$ dummy variables are needed to cover all categories.

Econometrics   The literal translation is "economic measurement." Econometrics is a branch of study that uses regression analysis to test theoretical models in economics.

Error Term   The difference between the *actual* value of an observation minus the *predicted* value for that observation. The error term captures all factors, including purely random ones, that a regression model has failed to take into account. *See* **Residual**.

F Statistic   A statistic used to perform an *F* test of significance.

F Test of Significance   In a multiple regression model, the *F* test of significance tests the hypothesis that all coefficients to the independent (*X*) variables are simultaneously equal to 0. That is, the *F* test of significance tests for the statistical significance of the regression *as a whole*. (This test can also be performed for subsets of independent variables and for nonzero hypotheses.)

Forecast   Using a sample regression model to predict the value of the dependent variable for a given value of the independent variable(s).

Gauss–Markov Theorem   A theorem that proves that if the linear regression model assumptions are satisfied (see Chapter 2), then the ordinary least-squares estimator has the least variance as compared to any other linear unbiased estimator (i.e., OLS is BLUE).

Glejser Test

A test for nonconstant error variance. The first step involves performing an OLS regression and saving the residuals. The second step is to run another OLS regression where the absolute value of the errors from the first regression is the dependent variable and one of the original model's X variables (the one we suspect is the source of the nonconstant error variance) is the independent variable. If the coefficient to the X variable in this second regression is statistically different from 0, then we have evidence of nonconstant error variance.

Heteroscedastic Errors

Also known as nonconstant error variance. The case where the variance of the error terms *is not* constant throughout the population regression line. *See* **Homoscedastic errors.**

Homoscedastic Errors

The case where the variance of the error terms *is* constant throughout the population regression line. *See* **Heteroscedastic errors.**

Independent Variable(s)

In a regression model, the variable(s) that determine, in part, the value of the dependent variable. Also known as "X" or "right-hand side" or "predictor" variables.

Interaction Variable(s)

Independent variables that are created by multiplying two or more existing independent variables.

GLOSSARY ▶

Linear
A regression function is considered linear if all the coefficients (i.e., the β's of a population regression function) enter the equation with the power 1.

Model Specification Error
The case where the regression model is improperly specified. Model misspecification can occur in a number of ways such as omitting a variable from the model that theoretically should be included, including a variable that theoretically should not be included, or using the wrong functional form for the model.

Multicollinearity
A problem in which one of the independent variables is related to one or more of the other independent variables. The extreme case of *perfect multicollinearity* occurs when one of the independent variables is an exact linear function of one or more of the independent variables. In this extreme case, ordinary least-squares estimates of the regression model coefficients are not possible.

Multiple Regression
A regression model with more than one independent variable.

Nonconstant Error Variance
See **Heteroscedastic errors.**

Normality Assumption
In the context of regression models, it is the assumption that the stochastic error term, u_i, follows a normal probability distribution function.

Null Hypothesis
Used in hypothesis testing, this is the hypothesis that is typically the focus of the test. It is tested

against the "alternative" hypothesis. *See* **Alternative hypothesis.**

Ordinary Least Squares (OLS)

A method for estimating the coefficients of a sample regression model. The coefficients are chosen such that the sum of the squared residuals is minimized.

P **Value**

Used in hypothesis testing to test whether a sample estimate of a coefficient is statistically different from 0. It represents the probability of achieving the estimated coefficient for the sample at hand, if, in fact, the population's coefficient is 0. If the *P* value is less than the chosen level of significance, then the zero hypothesis is rejected.

Pooled Data Set

A data set in which a cross section of data is followed over time. For example, a data set containing figures for each of the 50 U.S. states over several time periods. *See* **Cross-section data set; Times series data set.**

Population Regression Function

A functional relationship between the dependent and independent variable(s) for a population (as opposed to a sample) data set. *See* **Sample regression function.**

R^2 **(R-Squared)**

Equal to the proportion of the variation in the dependent variable explained by a sample regression model. It is a measure of the overall goodness of fit for a sample regression function where values close to 1 represent a "good" fit and values close to 0 represent a "poor" fit.

Residual   The difference between the *actual* value of an observation minus the *predicted* value for that observation in a sample regression. The residual (denoted *e* in this book) is similar to the error term (denoted *u*), except the former is typically used for the sample regression function, whereas the latter is used for the population regression function. (Note: In this book, both terms are used more or less interchangeably, where the context of the discussion determines whether they are population or sample regression function errors.) *See* **Error term.**

Sample   The functional relationship between the dependent variable and one or more independent variables for a sample of data (as opposed to that of the population data). *See* **Population regression function.**
Regression
Function

Significance   Same as *P* value.
("Sig.")

Significance *F*   The significance *F* shows the probability of obtaining the estimated values of the *X* coefficients from a *sample* regression if it were true that all the *population*'s *X* coefficients are simultaneously equal to 0. If the significance *F* is less than our chosen level of significance, then we can conclude that the model *as a whole* is statistically significant in explaining the values of the dependent variable. *See* ***F* test of significance.**

Standard Error   Equal to the positive square root of the variance. A measure of the typical difference between the value of a random variable and the random variable's mean. *See* **Variance.**

Stochastic   Variables or processes that are inherently random (i.e., not deterministic or exact).

_t_ Statistic   A statistic that can be used to tests hypotheses over estimated coefficients from a sample of data. Typically used to test the significance of separate estimated X coefficients. Generally, the larger the _t_ value, the more likely the estimated coefficient is statistically important. *See* **P value; _t_ value.**

_t_ Value   Used in conjunction with a _t_ statistic to conduct tests of hypotheses for sample data results. Given a chosen level of significance, if the absolute value of the _t_ statistic is greater than the given _t_ value (obtained from a table of _t_ values), then the hypothesis under consideration (i.e., the null hypothesis) can be rejected. *See* **_t_ statistic.**

Time Index   A variable created for use in estimating time trends. A time index assigns a number to each period's observation of a dependent variable. The time index can then be used as an independent variable in a regression, explaining the values of the dependent variable across time.

Time Series Data Set   A data set that records values of a variable across time. For example, monthly sales figures for a single firm. *See* **Cross-section data set; Pooled data set.**

Trend   An estimate of how a dependent variable tends to behave over time. A *positive* trend suggests that a dependent variable tends to increase over time; a *negative* trend shows that a dependent variable tends to decrease over time. Can be used in forecasting. *See* **Forecast.**

Two-Variable Linear Regression Model	A linear regression model that has an intercept term and one slope coefficient (i.e., only one independent variable). *See* **Multiple regression**.
Unbiased	If, in repeated sampling, the mean value of a sample estimate is equal to a population's parameter, then the sample estimate is said to be an unbiased estimator of the population's parameter.
Variance	A measure of the dispersion or spread of a random variable around its mean. For a sample of data, it is equal to the sum of squared deviations for a random variable from its mean, divided by the degrees of freedom.
Weighted Least Squares	A regression method used to correct for nonconstant error variance. The method first weights each observation on the dependent and independent variables according to their variance, then a least-squares estimation is performed on the weighted data. *See* **Nonconstant error variance**.

REFERENCES

Aldrich, J. H., & Nelson, F. D. (1984). *Linear probability, logit, and probit models.* Newbury Park, CA: Sage.

Berry, W. D. (1993). *Understanding regression assumptions.* Newbury Park, CA: Sage.

Berk, K. N., & Carey, P. (2000). *Data analysis with Microsoft Excel.* Pacific Grove, CA: Duxbury.

Berry, W. D. (1993). *Understanding regression assumptions.* Newbury Park, CA: Sage.

Cochrane, D., & Orcutt, G. H. (1994). Application of least squares regressions to relationships containing autocorrelated error terms. *Journal of the American Statistical Association, 44,* 32–61.

DeMaris, A. (1992). *Logit modeling: Practical applications.* Newbury Park, CA: Sage.

Diebold, F. X. (1998). *Elements of forecasting.* Cincinnati, OH: South-Western.

Einspruch, E. L. (1998). *An introductory guide to SPSS for Windows.* Thousand Oaks, CA: Sage.

Fair, R. C. (1996). Econometrics and presidential elections. *Journal of Economic Perspectives,* Summer, 89–102.

Geary, R. C. (1970). Relative efficiency of count of sign changes for assessing residual autoregression in least squares regression. *Biometrika, 57,* 123–127.

Glejser, H. (1969). A new test for heteroskedasticity. *Journal of the American Statistical Association, 64,* 316–323.

Greene, W. H. (2000). *Econometric analysis* (4th ed.). Englewood Cliffs NJ: Prentice Hall.

Gujarati, D. N. (1995). *Basic econometrics* (3rd ed.). New York: McGraw-Hill.

Hardy, M. A. (1993). *Regression with dummy variables.* Newbury Park, CA: Sage.

Hsiao, C. (1990). *Analysis of panel data.* Cambridge, UK: Cambridge University Press.

Judge, G. C. H., Griffiths, W., & Lee, T. (1985). *The theory and practice of econometrics.* New York: John Wiley.

Kahane, L. H. (2000). Antiabortion activities and the market for abortion services. *American Journal of Economics and Sociology, 59*(3), 463–480.

Kramer, G. H. (1971). Short-term fluctuations in U.S. voting behavior, 1896–1964. *American Political Science Review, 65,* 131–143.

Liao, T. F. (1994). *Interpreting probability models: Logit, probit, and other generalized linear models*. Newbury Park, CA: Sage.

Medoff, M. H. (1988). An economic analysis of the demand for abortions. *Economic Inquiry*, 26, 353–359.

Menard, S. (1991). *Longitudinal research*. Newbury Park, CA: Sage.

Menard, S. (1995). *Applied logistic regression analysis*. Newbury Park, CA: Sage.

Scully, G. W. (1974). Pay and performance in Major League Baseball. *American Economic Review*, December, 915–930.

Stigler, G. J. (1973). General economic conditions and national elections. *American Economic Review*, May, 160–167.

Thorn, J., & Palmer, P. (Eds.). (1997). *Total baseball: The official encyclopedia of Major League Baseball*. New York: Viking.

Zimbalist, A. (1992). Salaries and performance: Beyond the Scully model. In P. M. Sommers (Ed.), *Diamonds are forever: The business of baseball*. Washington, DC: Brookings Institution.

INDEX

About the Author

Leo H. Kahane is Associate Professor of Economics at California State University, Hayward. He holds a B.A. degree from the University of California at Berkeley (1985) and a Ph.D. from Columbia University (1991). He is the author of numerous journal articles published in *Economic Inquiry, Applied Economics, Public Choice, Atlantic Economic Review* and elsewhere. He is also the co-founder and co-editor of the *Journal of Sports Economics*.

e-mail: lkahane@csuhayward.edu